Practical House, Wa
Automobile Pai

Including sign painting, ar
hints and recipes

W. F. White

Alpha Editions

This edition published in 2024

ISBN 9789361470813

Design and Setting By

Alpha Editions
www.alphaedis.com

Email - info@alphaedis.com

Contents

PRACTICAL HOUSE PAINTER.

The following is an infallible and simple commercial test of the purity of white lead:

"Take a piece of firm, close-grained charcoal, and near one end of it scoop out a cavity about half an inch in diameter and a quarter of an inch in depth. Place in the cavity a sample of the lead to be tested, about the size of a small pea, and apply to it continuously the *blue* or *hottest* part of the flame of the blow-pipe; if the sample be strictly pure it will, in a very short time, say in two minutes, be reduced to metallic lead, leaving no residue; but if it be adulterated, even to the extent of ten per cent. only, with oxide of zinc, sulphate of baryta, whiting or any other carbonate of lime (which substances are the principal adulterations used) or if it be composed entirely of these materials, as is sometimes the case with cheap lead (so-called), it cannot be reduced, but will remain on the charcoal in an infusible mass.

"A blow-pipe can be obtained from any jeweler at small cost. An alcohol lamp, star candle, or a lard oil lamp furnishes the best flame for use of the blow-pipe. This test is very simple and any one can very soon learn to make it with ease and skill."

JAPAN.

Always cut your japan in a little turps before you add it to the paint. An ounce of japan, cut with turps, will do better work than two ounces in *oil paint*, if put in clear. *Don't add dryer to any more paint than you can use up in a few hours*, because it will soon commence to fatten your paint in the pot and lessen its covering and wearing properties. Many a job has been spoiled by using old color, doped with japan. Such paint is liable to mildew.

OBJECTIONS TO THE USE OF CHEAP DRY OCHRE AS A PRIMER.

1st. It is too dark for light colored work, because sooner or later it will show through in spots, or darken the entire work.

2d. It leaves a rough, coarse surface which the succeeding coats fail to completely level up.

3d. Succeeding coats are liable to scale from cheap coarse ochre priming.

PAINTING TIN ROOFS.

When paint scales from a tin roof it is not always the fault of the paint. It stands the painter in hand to carefully examine a new tin roof before painting it. When the tinner uses rosin as a flux to make his solder flow, the rosin is melted and cools again on the tin. When such is the case, carefully scrape it off with a knife, otherwise it will be liable to scale off, and take the paint with it.

When acid is used in the place of rosin it is apt to corrode the tin, hence it is best, if you want a permanent job, to clean off the acid. To do this, first rub the seams with kerosene oil, then wash with soap suds and rinse with clean water. If the roof is quite new, and the tin feels greasy, go over it with a wash made of one pound of sal-soda to six quarts of water, let it stand one-half day; then wash the tin with clear water.

Instead of this method, I have given new tin a good rubbing with No. 1 sandpaper to make it hold the paint.

ANALYSIS OF OCHRE.

Below is an analysis of a sample of French ochre, which is about the average of that pigment:

	Parts.
Hydrated oxide of iron	42
Alumina	20
Silica	38

The oxide gives the color; the parts as given above are in the right proportion to give the most stable color and durable body to be found in ochre.

Here is an ochre, which was ground in a linseed oil substitute, and sold to the trade at four cents per pound in twenty-five pound cans, and retailed to the painter at *seven cents per pound* in cans, to-wit:

	Parts.
Barytes	58
Whiting	15
Oxide of iron, silicate and alumina	24½
Chrome yellow	2½

This so-called ochre could be ground in one-half the oil it would take to grind yellow ochre.

ANOTHER.

	Parts.
Poor chrome yellow	8
Ochre	25
Whiting	67

Ground in snide oil, and sold to jobbers at five cents per pound, to painters *eight and ten cents.*

ANOTHER.

	Parts.
Barytes	62.90
Ochre	40.00

Barytes is not ochre, and this was *sold as pure ochre.*

ANOTHER.

Sold as French ochre, and recommended for priming:

	Parts.
Oxide of iron, alumina	19.79
Silica	40.93
Whiting	11.57
Barytes	26.64

ANOTHER IN OIL.

	Parts.
Chrome yellow	12
Whiting	25
Barytes	63
Oil	13

The markets are flooded with such imitations of ochre, both dry and in oil. The quantity of oil required to grind pure French ochre makes it high-priced, hence there is a motive for putting up barytes, which takes but little oil in grinding.

YELLOW IRON ORE.

Much of the so-called dry ochre on the market is a *yellow, iron ore and not yellow ochre.* When mixed in oil and put on a tin roof it will turn brown inside of ninety days. I presume you have had experience with such stuff. This makes a bad primer; it is very liable to scale.

CRAWLING PAINT.

When paint crawls it is because there is not sufficient adhesion between the undercoat and the new coat, caused usually by too much gloss on the undercoat. To prevent crawling subdue the gloss on the undercoat by sandpapering, rubbing, or by the application of some material which will have the desired effect; or, if on the outside, wait until the gloss has been subdued by the elements. There is nothing more trying to one's patience than to have the paint let go and crawl up in bunches after it has been carefully brushed out. Hence, it is well to provide against such trouble in advance. The observing painter has no doubt noticed that paint is more liable to crawl under cornices, and upon other sheltered positions, than elsewhere; hence, it is best in all *such sheltered places, where the elements do not have full play, to use sufficient turpentine to prevent a high gloss on the undercoats.*

TO PAINT BLINDS AND NOT DAUB YOUR HANDS.

First, have a stick to open and shut the slats with after you commence to paint. Second, leave a place on each stile, or side rail, half way between the hinges, six or eight inches long, unpainted, except to cut in the edge next the end of the slats to take hold of when you turn the blind over or set it aside; also leave the bottom hinge unpainted. After you have set up the blind hold it up by the unpainted hinge until you finish the stiles; then lean it up against its support and touch up the hinge. In this way you need get no more paint on your hands than you would in painting a door. No time will be lost, because you can touch up the stiles in less time than it would take to wipe your hands and brush handle.

LEGLESS STEP-LADDERS.

Step-ladders without legs for outside work are good things to have on the job. Say, three of them, 6, 8 and 10 feet long. A man of good height can paint 14 feet high from the 10-foot ladder. They are much easier on the feet than a "round" ladder. You can stand straighter and reach farther when standing on a step than you can while trying to balance on a round stick; besides, a step is a handy place to set your pail on.

SUCCESS IN PAINTING.

Painting don't pay, eh? No wonder it don't pay, because here you are spending half your time growling. The facts in the case are, "You are not up-to-date." If there is no possibility of making money at the trade, how is it that your competitor gets along so well? Why is it that he accumulates and you lose? He goes into the same market for labor, material and jobs that you do. He comes out every fall with his pockets full, and you round up poor as a church mouse. There must be a screw loose somewhere in your management. Will I point one out? Certainly, we have always been friends, and I can never do too much for a friend. In the first place you are too impetuous. You forget for the time that bills for labor and material will fall due, that you must live—and you take the job at losing figures. You ought to realize that the success of a contracting painter depends upon his business qualifications. To-wit: Correct and careful estimates, coolness in bidding, care in selecting materials and men, systematizing his work so as to keep each man in the right place. I don't know how much you are getting for this job, but it looks to me that you are losing money every day by using poor material and improper handling of your men. The good business man prefers the strictly pure Dutch process white lead to the adulterated brands. He uses pure linseed oil instead of adulterated mixtures and imitations of it, and he never loses sight of the fact that a good reputation is a mine of gold to him. If he finds a man is a good hand on a ladder or swing stage he keeps him there, and if he finds a man an expert at inside work he keeps him there, and if he finds a man is a poor stick in any place he lets him go, rush or no rush. If he has high work he provides a safe and easy way to get there. If he has inside work his step-ladders are equal to the work. He knows when a man has to reach too far or stand on top of a ladder he can't half work. Learn to manage your men, to keep the right man in the right place. Stop making ruinous bids. Open your eyes to the fact that a man who makes a losing bid on a job, to beat his competitor, acts like an idiot, and is meaner than flies in paint.

GUESSING ON WORK.

The practice of estimating work by guess has brought many a painter up with a round turn in the fall in debt. The curious part of it is that the lesson is rarely, if ever, learned. Don't be too smart. Guessing on work is very uncertain business.

GLASS GILDING.

A practical expert in an English journal, the "Plumber and Decorator," gives the following as his process acquired and tested by many years' experience.

The tools and materials required for glass gilding are the same as used for gilding in oil, excepting the gold size. Oil gold size would never do for glass work. In glass gilding the object is to get a size or mordant which will have the least possible tendency to destroy or mar the burnish of the gold leaf. This is absolutely necessary, when we consider that in this kind of work the size is before the gold, not as in oil gilding—behind it. For a mordant nothing can be better than the best isinglass. To prepare this for use the utmost care and cleanliness should be exercised. The water must be quite pure—free from grease or impurities of any kind. In preparing the size the following may be relied upon as a first-class recipe: Boil about one pint of water in a perfectly clean pan. Should any scum rise during the operation remove it with a large spoon. Then add about as much isinglass as will lie on a dime to the boiling water. This is best done a little at a time to prevent it gathering in a mass before it has a chance of dissolving. When the isinglass is dissolved strain the size through a fine silk handkerchief, folded double or fourfold, or, better still, through some white blotting paper. This straining or filtering will remove any bits or impurities that may have lodged unperceived in the isinglass. When cool the mordant is ready for applying to the glass. Some gilders like to add spirit in some form—generally spirits of wine—to their size. Their reasons for doing this are not always very explicit. Some do it because they have seen others do it. Others add it, they say, to give the gold a better burnish, or to make it better adhere to the glass. This is a delusion. The most sensible reason for its use was imparted to me by a veteran in the trade. He used spirits of wine to take out or kill any slight greasiness that may have been in the water or isinglass. I must confess that until I learned this, spirits always formed part of my mordant, because others used it. However, on further consideration, its use has been discarded, and, if anything, a better burnish on the gold is the result. In making the size it must be borne in mind that the less isinglass used the brighter will be the gilding when completed. Of course, if too little be used, the gold will not adhere to the glass as it should, and this would cause much damage and annoyance when the isinglass size was floated on again to proceed with the second gilding. When the size is too strong, or contains too much isinglass, no amount of burnishing will remove it altogether from before the gold. These are important points and should be carefully studied. But a little practice soon teaches the gilder how to arrive at the happy medium.

There are a variety of purposes to which ornamental glass gilding may be applied besides sign work, shop fronts or glass doors. It is now much used for show cases, window tablets, druggists' bottles, fixtures and pilasters for shop fronts. Very often the design is embossed or bit into the glass, and worked up with gold and silver leaf, besides being picked out in colors. This is both a costly and effective method of decorating, which shall have full consideration in a future chapter. For the present it will, no doubt, be advisable to consider the simpler form of glass gilding. When this is thoroughly understood very little further instruction is needed for high-class work.

For the sake of example we will suppose a glass slab about three feet six by twelve inches is the subject to be treated. This is to have black letters without thickness or shadow on a gold ground. There are two methods of doing this. One is to first paint on the glass the letters with japan black and afterwards gild the plate. The other consists in first gilding the plate solid and then painting in the background with japan black. By this method the lettering is left untouched. The gold on these is then washed off, the edges trimmed, and the letters themselves painted black or any other desired color. This latter is, perhaps, the most satisfactory. However, a few lines of explanation will be devoted to each process.

First in order comes a plate, the letters on which are painted with japan black previous to gilding. To the learner, no doubt, the plain block letters will prove an attraction, because of their simplicity. This should be set out correctly on a sheet of lining paper. It will only be necessary to run in an outline of the letters. When completed to the satisfaction of the operator it may be pasted round the edges and fixed on the face of the glass. The back of the glass, that is the side upon which the work is done, should be quite clean. When the plate is fixed on an easel or stand, which is the most convenient place for working, the letters will, of course, read backwards. In this form they must be painted. When quite dry and hard, should the outlines of the letters be irregular, they may be set right in a very simple manner. All that is required to accomplish this is a metal straightedge and a sharp quarter inch joiner's chisel. The straightedge is laid across the tops and bottoms of the letters and the chisel is employed to cut them sharp and true. The sides of the straight letters are then similarly treated; curves must be perfected with a writing-pencil. The paper may now be taken from the face of the glass and the plate examined all over. Should any specks of black be found on it they must be removed before the gilding is gone on with. The smallest speck shows up before the gold leaf. If convenient, before gilding, have the plate fixed at an angle of about 45 degrees. While in this position take a flat gilder's mop and float the isinglass size over the glass. Then take up the cushion, take out a few leaves of gold from the book, and whilst the glass is wet cover it with gold, lifting the gold from the cushion as described in the last chapter. If possible lift a whole leaf at once, but should this at first prove troublesome try half a leaf. Keep the glass wet with the size and overlap slightly each successive leaf of gold until the whole of the glass is covered.

The glass must now dry before it can be re-gilded, and must then be gently rubbed with the finest cotton wool. It is an easy matter to ascertain whether it is dry or not. When wet the gold, if looked at from the front of the glass, has a dead look, but if dry it shows up bright. If possible leave the plate till next day before giving it a second coat of gold. The advantage of this delay is obvious. The gold has time to get hard, which materially reduces the risk of its being removed when a second application of size is necessary.

To the novice it will, no doubt, appear at first sight both a waste of time and gold to gild all over the work again, but if the plate be held up to the light it will show many imperfections in the shape of small holes, cracks and imperfect joinings. Another coat of size floated on and another layer of gold over the whole of the work should turn out a so far satisfactory finish. Let the glass dry again and be subjected to a further examination for faulty places. Should any be found cover them with more gold. But if the work is satisfactory it is ready for the burnishing process. The first stage is to polish the gold by gently rubbing with fine cotton wool, care being taken not to scratch the gold. This is, of course, only a repetition of the polishing after the first layer of gold.

There are several methods in general use for obtaining that brilliant burnish so much admired in glass gilding. But the one that meets with most favor and success is what is called the "hot water burnish." It will be advisable to practice on the glass under consideration. After the cotton wool polishing is completed warm the glass either by holding it before a fire or gently pouring warm water over it. This is only a precaution against breakage by sudden expansion. Now let it dry, and while warm polish again with the cotton wool. Repeat the pouring of water, hotter than the last, and when the glass is dry, after this operation, gently rub it again with cotton wool. This hot water flushing should be carried on until the burnish is quite satisfactory. But it must be very carefully done, else the gold runs a risk of being washed off in patches.

HOW TO ESTIMATE WORK.

Measure your work with the tape-line and be sure you get all there is in it; projections, depressions, mouldings, edges, etc.

Many a painter has dropped his profits by not taking in these little particulars. Every bead, sunken or raised panel makes an edge to paint. The edges of ordinary weather boarding *add ten per cent.* to the surface, to say nothing of the edges of the corner boards and window and door casings—the projections and depressions in the panels of an ordinary four-paneled door, *add at least ten per cent. to the surface* to be painted. Then let me say to you again, look closely for edges, projections, depressions, hollows and rounds. They all count when you paint them; and it is your fault if they are not included in the estimate. When you have multiplied the number of feet around a house by the average height and reduced it to yards you have only made a start. Measure the cornice, follow the hollows, rounds and edges with the line. *There is lots of surface in mouldings.* The tape-line is good as far as you can make it go, but it can't do it all. You must use judgment in connection with it; and carefully estimate the condition of the work, what per cent. is *slow* to paint, or high and difficult to reach. For instance, what is the condition of the surface, is it porous and full of cracks? Is every joint gaping for putty? Is the putty on the windows rough and broken? Is the old paint cracked, blistered and scaling? Is the cornice ornamented with dentils, brackets and panels? You may lose a day or a week of extra time on a high tower or cupola if you fail to put it into your estimate as extra hard to reach. Make the price accordingly. Are the blind-slats stuck fast and difficult to paint? Is the work to be done in the busy season when labor and material are high priced and good men are hard to get; or in the dull season, when dealers will cut prices and good men are hunting for work? Bidding on specifications must be done with care. You can figure the number of yards to be painted, but there are many points which the completed job can alone disclose. A provision in your contract to cover all changes in specifications comes mighty handy on the day of final settlement. It is not safe to make anything like a close bid on specifications, until the following questions have been settled and put in your contract. To-wit: Will the building be delivered to you at a specified time, finished and *cleaned out* and put in good condition for the painter; or will you be expected to commence before the work is finished and paint as the work is put up, and spend as much time dusting and sweeping as you do at painting? Will the machine-dressed lumber, including mouldings, doors, window-stops, etc., be put in as it comes from the factory rough and fuzzy, or will it be redressed and made smooth and ready for the paint? These points may look to you like small matters, but they count when you come to paint the work. If you are to do a fine job stipulate in your contract that the wood-work, etc., shall be finished in good shape. If you are to paint the work as you find it have it so stated in your contract. Paste this motto in the top of your hat and read it often: "It is always better to lose a job than to get it and lose money on it."

Two houses may be of equal dimensions, yet it may be worth 50 per cent. more to paint one than the other; hence any definite scale of prices for work by the yard is liable to be misleading. We may determine by the line how much there is of the work, but we must rely upon our judgment and experience to determine how much it will cost to do it.

ONE WAY TO MEASURE A JOB.

Find the surface measure of the entire job, including all edges and projections, and estimate how much it is worth per yard, on the basis that it is all plain work, easy to get at. Next we will proceed by what we may call special measurement. Suppose the cornice measures 60 yards, and is finished with blocks, moulded panels and brackets, and we estimate that the cost of painting it will be three times that of a plain cornice, hence we will add two measures or 120 yards to the general or first measurement. Next, suppose each window and casing measures three yards, and there are 20 of them to be trimmed in colors, we estimate the work of painting them double that of plain work; hence we add to our special measurement 60 yards. If there is a cupola high and difficult to reach we estimate that it will be worth double the cost of painting ordinary work to do it. Say, it measures 50 yards, we will add 50 yards to the general measurement, and so we will go on until we have taken in all parts of the work which will cost more than ordinary plain work.

To illustrate: The building measures 600 yards, and as plain work we estimate it worth twenty cents per yard to paint it. We amount our special measurement which we will say adds up to 300 yards, which added to the 600 yards general measurement, makes 900, which at twenty cents per yard, makes $180. The same system may be used inside.

TO SOFTEN HARD, LUMPY PUTTY.

Break the putty into lumps; put it in a kettle with enough water to cover it; add a little raw oil, and boil and stir well while hot. The putty will absorb the oil; pour off the water, let the putty cool, then work it, and your putty will be as good as new.

TACKY PAINT ON CHURCH SEATS, ETC.

During my experience as a painter, I have been called upon to repaint tacky seats in at least half a dozen churches. Such seats are an unmitigated nuisance. Tacky paint may be the result of putting too much japan in oil paint, or of using fat oil, or paint which had been mixed a long time, especially if it had very much japan in it, or by mixing oil and varnish, or by putting varnish on oil paint, especially if the paint had not been given time to dry hard before it was varnished. To harden tacky paint try this: Take one part japan and three parts of turpentine, and give the work a coat of the mixture. That will usually effect a cure, unless the paint is soft clear to the wood. A coat of shellac will sometimes do the work all right. Such seats usually seem all right until warmed by the heat of the body; hence we may be satisfied that the fault is in the oil used in the paint or varnish. It is best on that account to use but little if any oil when painting seats of any kind. Coat up with color ground in japan and thinned with turps; varnish the part which comes in contact with the body with shellac varnish.

I have painted seats this way, and never heard of any further trouble with them.

To repaint tacky seats the best way is to burn off the old paint, and coat up as above; because, if a hard drying paint is put over the old soft paint it is liable to crack. It is well, however, to see if the turpentine and japan will work a cure, or if a coat of shellac will stop the trouble. To do this it is well to first experiment on one seat, or upon a small surface.

I have killed tacky paint by rubbing it with a cloth wet with ammonia; when dry, try it, and see if the "tack" is gone; if not, go over it again; when dry, put on a coat of shellac varnish; this is a pretty sure cure.

TESTING JAPAN.

If japan smells of benzine don't buy it. Mix it with clear oil; if it curdles, you don't want it. Mix drop black with some of it; as stiff as good drop black ground in japan; then thin with turps and make a painting test, to see if it is a good binder. To see if it will crack, paint on glass, let it dry and hold the glass between your eye and the light. If you see fine cracks don't buy any of it.

When you go to buy japan, ask the dealer who made it. If he don't know, make up your mind at once that it is a *fatherless waif* without *a name*, and likely to be worthless. When a man makes a good thing he is apt to send his name along with it as an advertisement. This applies to all material. There is a great deal of bad japan on the market, and a great amount of work ruined by it. Buy none unless it bears the brand of a reputable maker and will stand these tests.

I do not need to tell the practical painter that there is a great amount of bad japan on the market, and that a great deal of paint is ruined by it. Buy no japan unless the can bears the name of some reputable manufacturer, and will stand the above tests.

WHY DO PAINTS AND VARNISHES CRACK?

The following paper was read by Mr. A. P. Sweet, of Iona, Mich., at a meeting of master car painters:

SUBJECT:

"Why do paints and varnishes crack, and what is the reason that cracks in the latter are usually at right angles to the grain of the wood?"

The subject, as I understand it, relates to the cracking of varnishes, etc., as experienced in connection with passenger car work, and as such I introduce it for discussion before this association.

There are many theories as to the cause of the cracking of paints and varnishes. Some are well defined, others are not satisfactorily explained.

I do not anticipate being able to add much to what is already known, but will advance a few thoughts, which may call forth the views of others on the subject.

The old adage, "It takes two to make a quarrel," is as true when applied to paints and varnishes as it is to individuals. A single coat of either seldom, if ever, produces cracks. These make their appearance only after two or more coats have been applied; consequently, it is necessary to have a body of color or varnish, consisting of two or more coats, before any trouble of this kind makes itself manifest.

This being the case, it follows that the cause of the difficulty must be sought for in the coatings themselves, either in the quality of the material employed or in the mode of applying them.

Poor and cheap oils and japans—especially the latter—are a fruitful source of cracking in paint; but by far the most prolific one, in my opinion, is the hurried application of the succeeding coats before the preceding ones are dry enough to receive them. If sufficient time is not given, cracks will inevitably follow such a mode of procedure.

I am of the opinion, also, that very little blame can be attached to the wood used in the construction of cars, as most of it is comparatively well seasoned, and its expansive and contractive force is not sufficient to cause serious trouble. If green wood was used there might be room for this excuse, especially where the cracks run in the direction of the grain, and are large and deep.

Before pursuing this subject further, it may be well to examine a little into the theory of the drying of paint. It is purely a chemical process, not a mechanical one, as some suppose. Paint dries by the evaporation of its volatile parts and its absorption of oxygen; it is heavier when dried than when in the liquid form, having attached to itself a sufficient amount of oxygen to very perceptibly increase the weight some 6 per cent.

The best grades of linseed oil are said to contain from 70 to 80 per cent. of substance called linoleine, a resinous and slow-drying oil and acid which imparts to the oil its elasticity.

In the process of drying, contraction occurs. The various atoms of which the coatings are composed move closer and closer together; and as this contracting force is easier with than across the grain, cracks at right angles to it are formed. This fact suggests the necessity of so adjusting the elasticity of the various coats that the force exerted in drying may be as nearly equalized as possible, as their contracting force is continued until all elasticity has left the paint and oxygen ceases to be absorbed, all the oil acid has disappeared, and nothing but a hard, brittle surface remains.

Under the microscope, in the first stage of cracking, the surface presents nothing unusual except that the cracks appear clean cut and sharp on the edges. As months pass by and the surface is exposed to the atmospheric changes of heat and cold, wet and dry, the cracks become more numerous; and in the last stage, when the oil is entirely destroyed, the surface assumes the appearance of innumerable rectangular masses, higher in the center than at the edges, like small mounds raised by the process of contraction and adhesion.

Cracking in color coats may, by careful attention to preliminaries, be reduced to a minimum, provided good first-class materials are used and sufficient time is given to each coat to dry.

Where varnish is to be applied as a finish, all coatings should have oil in their composition and yet be mixed to dry flat. They should be applied very evenly and thinly, even if it necessitates an extra coat, to cover and make a solid job.

Striping and ornamenting should be done on flat color, which gives time for hardening, and fits it for the varnish coats to follow. If work is done in this way, I think very little fear of premature cracking need be entertained; at least, not until time and weather have sufficient opportunity to play havoc with its beauty, and natural decay of the materials themselves necessitates a thorough overhauling and repairing.

Rubbing varnishes are another source of trouble, causing the succeeding coats of finishing varnish to show signs of cracking long before they otherwise would, as it does not agree with the slower drying varnishes usually applied above it, being of a harder and more brittle character, serving the purpose of producing a fine, smooth surface, but sacrificing the durability of the job.

Concerning the cracking of varnish, I have not much to say. It seems to me that many of the reasons given above will apply to it as well as to the paint.

Poor material in the shape of varnish is poor indeed. A first-class article only will give first-class results.

It must be elastic, or it will crack easily and badly, no matter how good the undercoats of paint may be.

Good varnish on good color coats will not give any signs of cracking until, by repeated varnishings, it has accumulated a thick coating of brittle, unelastic gum.

No painter can say truthfully that his cars never crack, as it is a natural consequence of decay, and will come, sooner or later, to the best of material.

That varnish cracks to a great extent at right angles to the grain of the wood, I think is due, in some degree, to the same reasons as given above for the cracking of paint, and after its elasticity is destroyed by age. Vibration has a great effect upon the hard and brittle coating of gum that remains, coupled with expansion and contraction caused by variations of temperature and the disintegrating influences of the weather.

BRUSH CLEANING TROUGH.

To make such a trough, take a piece of planed board, 6 inches wide and 18 inches long, and nail on side pieces 2 inches wide; this makes the trough. Nail this trough on a bench, box, or table, and let one end of it project over the edge of the bench, box or table, and place your slush bucket under the projecting end of the trough. To clean a brush, lay it in the trough, keep hold of the handle with one hand and with the other take a dull scraper and press the paint out of the brush and shove it off into the slush bucket. The advantage of this method is that you clean the whole length of the brush and save the paint, instead of daubing it on the walls of your shop.

FLOOR WAX.

A good preparation for waxing floors may be obtained as follows:

Yellow Wax	25	oz.
Yellow Ceresin	25	oz.
Burnt Sienna	5	oz.
Boiled Linseed Oil	1	oz.
Turpentine	1	gill

Melt the wax and ceresin at a gentle heat, then add the sienna previously well triturated with the boiled linseed oil, and mix well. When the mixture begins to cool add the oil of turpentine, or so much of it as is required to make a mass of the consistence of an ointment.

The burnt sienna may be used in smaller or larger quantity, according to the tint desired, or may be replaced by raw sienna, etc.

DAMAR VARNISH.

Never use damar varnish over oil paint.

Never put oil in damar varnish. See to it that your dealer does not draw it into an oil measure, and that you do not keep it in an oily or rancid can. Why? Because it is liable to dry tacky under any of the above conditions.

STENCIL STAINING.

Ordinary plain staining can be done by almost any one who can handle a common paint brush. Yet it is not generally known, even to skilled decorators, that stain, on sound white wood, evenly planed, can be applied to imitate the most intricate of artistic designs; such, however, is the case. A decorator if asked to imitate in stain on white wood a piece of parquetry or inlaid wood, might reply that such a thing was impossible, alleging as a reason that by employing liquid stain in the same way as a distemper—that is to say, by the aid of a stencil to reproduce the pattern—the stain, as soon as it became absorbed would be found to "run," and so giving to the pattern imitated an indistinct or blurred edge. Yet the most elaborate patterns are successfully stenciled direct on to pine, and the figured work on this wood has invariably come out distinctly and naturally as to be almost indistinguishable from the inlaid work they have so successfully sought to imitate. The great difficulty to be overcome in stenciling with stains is undoubtedly the "running," but with a very little care and patience this can be easily obviated. Say a painter has a border to stain round an ordinary pine floor in imitation of a selected pattern of parquetry, the colors of which are generally in two or more shades of oak, the first thing he has to do after having properly prepared the floor—namely, making the part to be stained as smooth and as even as possible by filling up the crevices and nail holes—is to stain over the work in the lightest shade shown in his pattern; this can be done by diluting the ordinary liquid oak stain with water to the desired tint. Next let him cut out of a piece of lining, paper in the form of a stencil—the pattern he has to reproduce on the floor—care being taken to oil the stencil in order to strengthen and preserve it. He should then mix the stain into a stiff paste or to the consistency of a distemper used for ordinary stenciling; place a portion of this mixture on a smooth piece of wood, take up a very small quantity of it on a stencil brush and apply through the stencil plate in the same way he would a distemper. If a very dark shade is required apply more stain before removing the stencil plate.

PAINTING BRICK.

Objections: Chipping of the brick, and scaling of the paint.

The chipping may be on account of defective brick or otherwise.

Scaling may be caused by poor paint, or by *dampness in the brick.*

When called upon to paint brick, first see if the brick is dry. See that there is no place where water leaks in from the roof or cornice and soaks into the brick. A brick wall may look dry and still be damp inside. If you want paint to stay on brick, give the brick time to dry, after heavy and driving rains. It is always a bad plan to paint brick in the fall, after the autumn rains. The only real safe time to paint a brick wall is in summer, after a spell of hot, dry weather. You can not always wait for that, but you can tell the owner that it is unsafe to paint a brick wall until it has had time to dry. Why? Because in winter the moisture, which is shut in by the paint, will freeze, expand and throw off the paint or chip the brick.

Prime brick work with a thin coat of good paint mixed in pure linseed oil. Flow on the priming freely, and brush it well into the brick; for second coat, whatever paint you use, put in at least one-fourth white lead; make this coat one-third turps, and rub it well out. Give it a good body. For the last coat, use your color regardless of lead, unless you want it in to get your color. If you want a gloss, mix this coat with all boiled oil, and flow on. For flat, if your colors are ground in oil, use one-fourth oil and three-fourths turps, and if it don't show flat when painted, it will flat in a short time. The last coat may admit of more oil or may not take as much, and flat. This depends upon the work when started, etc. Some painters make brick flating by breaking up the pigment in japan, and elastic varnish for a binder, and thin with turps. I prefer the oil for a binder, and have made the last coat one-half oil, and had a nice flat in a few weeks. I always ridicule the idea of painting brick flat, because it will not stand as long as an oil finish, and the oil finish will be flat enough in a few months.

CLEANING UP A ROOM.

Now, if I were going to teach a boy to clean up a room, the first thing would be how to prepare himself for the job. In the first place, he wants a damp sponge with a string through it to tie over his head, to hold the sponge over his mouth and under the nose to catch the dust, because it is a great deal more pleasant and a "sight" more healthful to carry lime and other dust in a sponge than in nostrils and windpipe. Then he wants a cotton cloth cap, large enough to draw down over his head and ears, bib overalls and jacket to button close about the neck and he is well fixed. In such a rig he may look peculiar, but he had better look like a monkey than to skin his nostrils with dust and fill his ears and hair with lime, sand and sawdust.

For tools, he needs a good, new, fine corn broom, a wide bristle sweeper (a ten or twelve-inch paper-hanger's smoothing brush will do), a good duster, a sharp tool to pick out corners, a two-inch chiseled brush for corners. A sprinkler only turns dust to mud, to dry in a few hours and become dust again. When you have swept the floor with your broom and dusted your wood-work and gone over the floor carefully with your wide bristle brush to take what you brushed from the casings and what the broom left on the floor, look at the air across this ray of sunlight; it is full of dust, soon the most of it will settle on the floor and casings and window stools. What then? Wait till it settles and *wipe it off with a cloth* and don't forget the tops of the doors and casings. "Why use a cloth?" Well, if you go in and begin to use a dust brush after the dust settles you throw a portion of it in the air again and it will settle on the work. And by the way, I want to say that a wiping cloth is a very important article for a painter to carry. It always makes me "red hot" to see a painter (?), after he has daubed a key shield or a hinge, try to wipe it off with his thumb; I could forgive him for the daub; the best man in the trade may sometimes do that, but the man who will rub part of it off with his thumb and let the rest dry ought to be sent off the job or suspended long enough to take a lesson in the art of wiping off daubs.

I want to say further that every well regulated dusting kit ought to have a dust pan hitched to it in some way. It will save sweeping the dust out on the steps to be tracked in again, save the time you would lose in sweeping the dust over thresholds, or save the time it would take to borrow one.

PASTE FOR LABELING ON TIN.

Make a stiff flour paste in the usual way, with flour and water, then add 2 ounces tartaric acid, and 1 pint of molasses; boil the mixture until stiff, and put in ten or fifteen drops carbolic acid.

ANOTHER.

Wheat flour	1	pound
Alum	2	drams
Borax	2	drams
Hydrochloric acid	1½	ounces

Mix the flour, alum and borax in the usual way, to a smooth paste in water, then add the acid and cook in the usual way with hot water.

TO MAKE TENTS, ETC., WEATHERPROOF.

To prevent tents, wagon covers, etc., from rotting dissolve 4 ounces sulphate of zinc in 10 gallons of water, then put in one-fourth pound sal-soda, stir well until dissolved and add one-fourth ounce tartaric acid. Let the cloth lie in this one day and night and hang up to dry. Don't wring it.

TO PAINT ON CANVAS OR MUSLIN WITHOUT SIZING.

First stretch, then wet the cloth. Wipe off the drops and letter while the cloth is damp with color mixed with japan and turps.

TO PAINT ON ZINC.

A difficulty is often experienced in causing oil colors to adhere to sheet zinc. Boettger recommends the employment of a mordant, so to speak, of the following composition: 1 part of chloride of copper, 1 of nitrate of copper and 1 of sal-ammoniac are to be dissolved in 64 parts of water, to which solution is to be added 1 part of commercial hydrochloric acid. The sheets of zinc are to be brushed over with this liquid, which gives them a deep black color; in the course of 12 to 24 hours they become dry, and to their now dirty gray surface a coat of any oil color will firmly adhere. Some sheets of zinc prepared in this way, and afterwards painted, have been found to withstand all the changes of winter and summer.

PAINTING BLINDS.

When painting a blind never turn it upon edge when cutting in the inside of the rail, because the paint will be likely to run into the pivot-holes and stick the slats. When you set a blind up to dry, set the bottom end up, and be sure to have the slats lie flat side up. Why? Because the bottom end of the blind when hung is more apt to drag on the window sill than the top end is to touch the jam above. If set bottom end up, that end will dry solid and if there are any sags it will be at the top. Keep the slats flat side up to avoid flat edges.

TREATMENT FOR HARDWOOD FLOORS.

First see that the floor is clean and smooth; then give it a coat of best oil, with japan sufficient to make it dry; cut the japan in turps. Then put on a good mineral paste, filler in the usual way by rubbing the filler well into the wood; then clean off all the surplus. When dry, sandpaper and putty up well with colored, hard putty, and put on a coat of shellac; if too glossy, rub down with powdered pumice and oil. Be careful to have the putty match the floor.

WHITEWASH FOR OUTSIDE WORK.

Take one-half pound of fresh burnt lime. Dip it in water and let it slack in the open air. Melt two ounces of bagundy pitch by gentle heat, in six ounces of linseed oil; then add two quarts of skim milk while the lime is hot, add the mixture of pitch and oil, a little at a time while hot, and stir it in; then add three pounds of bolted whiting and stir. Add more milk if too thick for the brush.

THE STRAINER.

Don't forget to use the strainer. After you have put in your best licks to clean up and sandpaper a job, it is the height of folly to daub it up with paint full of skins and specks. Oil paint is liable to be "skinny" in the keg. Miller's bolting cloth makes a good strainer, and common cheese cloth at five cents a yard does very well for ordinary purposes.

TO KILL GREASE SPOTS ON WOOD.

Use a wash of saltpeter or a thin lime wash, then rinse with clear water. Treat blacksmith's smoke in the same way.

KALSOMINE.

To please an old friend I give the following recipe for kalsomine. *He says it is good.* I never used it, so you will have to take his word for it.

Fifteen pounds good paris white, mixed up in lukewarm water, add one-fourth pound good glue, dissolved in the usual way, strain through a fine sieve, then dissolve one-fourth pound white hard soap in hot water and one-half pound of alum in cold water and mix. Add water to give the right consistency for putting it on the wall.

TO TAKE OFF THE PAINT.

If you have an old, roughly painted door to cut down for a fine job, don't fool away your time, and fill your nose with dust, trying to do it with dry sandpaper, but take the door off its hinges, lay it flat on horses, and keep the surface under your sandpaper wet with benzine, and you can do in an hour what would otherwise take half a day. The benzine softens the paint, and keeps the paper from gumming up. If it is not practicable to take the door off the hinges, put your benzine in a small spring-bottomed oil can and squirt it on the work as needed to keep the paper clear of paint and make it cut fast. Wipe off the loose paint with rags. It works equally well on old varnish. Try it once on an old carriage body.

If the old paint is extra hard use a mixture in equal parts of benzine and ammonia.

CLEANING SILVER, BRASS OR COPPER.

In the course of our work we often meet with tarnished metal ornaments, which must be cleaned to make our work look well.

This preparation is a good one:

Paris white (fine)	1	pound
Carb. magnesia	2	drams
Cyanuret potash	7	drams
Sulph. ether	3	drams
Crocus martis	1	dram
Soft water	1½	ounces

or sufficient to make a stiff paste.

Mix by rubbing, add the paris white last, then stir into the water. Apply with a rag or sponge, and rub dry and polish with a rag or canton flannel.

WHY DO WALL PAPERS CRACK?

Some papers are more inclined to crack than others, because they are made of more brittle material. When selecting a paper for a whitewashed wall or ceiling, take a pattern which feels soft and pliable. Papers which crackle or rattle when crumpled in the hand are liable to crack. Papers which stretch or expand the most when wet are the most apt to crack; because when they dry and shrink the pull is so great that the fibers give away, if great care is not taken in putting it on. Cracking may be the fault of the paper hanger. He may use his paste too thick, or too thin, or put on too much or too little. Paste should be put on even and of the proper consistency and thickness to cement the paper to the walls. Paper is more liable to crack on rough and uneven walls. On a smooth wall, if properly put on, it becomes, as it dries, so fastened to the plaster that it cannot contract enough to break the fibers, but on a rough and uneven wall there are apt to be loose places where the air gets in, and the contraction of the paper so weakens the fibers that it cracks.

Now, if the paper hanger will be careful to secure the paper uniformly by using sufficient paste on rough places to hold the paper, and be careful to brush or pound the paper down firmly, he will greatly reduce the chances of cracking. A roller can not be depended upon for a rough wall. Too much or not enough sizing on a wall may be a cause of cracking. Hot paste, which thickens as it cools, is not safe to use on such walls, because it may appear just right when hot but will be too thick when cool and cause the paper to crack.

OIL SIZE FOR WHITEWASH.

Oil size is good to use on a whitewashed ceiling before papering if you don't overdo it. A friend of mine thought, if a little was good, a great deal would be better; so he gave his ceiling two flowing coats of clear oil, and when dry put on his paper, but it did not stay. Why? Because he put on so much oil that he made a glossy surface *and the gloss could not hold the paste.* An oil size on whitewash is all right if used right. It is a mistake to use clear oil; 1 pint of oil, 1 pint japan and 1 quart turpentine is better, because it will penetrate further, dry faster, flat the surface, and have sufficient binding power to hold the whitewash from coming off. Don't size a wall with paste. Paste and whitewash don't go well together. The fact that you have to size your wall to make paper stick proves this.

Oil size should dry hard before the paper is put on.

I find glutol, manufactured by the Arabol Manufacturing Co., No. 13 Gold street, New York, a first-class substitute for glue in wall size and kalsomine, and prefer it to glue, because it will not attract flies, nor spoil by standing in hot weather, and can be mixed in cold water.

TO CLEAN BRICK.

The white powder which comes on brick can be removed by sponging with a mixture of muriatic acid and water, equal parts. Wash the brick in clear water and let them become well dried before painting.

TO CLEAN TARNISHED ZINC.

Mix 1 part sulphuric acid with 12 parts water and rub the zinc with it with a rag, then rinse with clear water.

TO GILD ON WOOD.

First get a good body and a smooth surface. The work should be flat with three coats at least on wood, and not less than two on iron or tin. The best size for outside work is oil gold size (fat oil), mixed with a little medium chrome yellow toned down with white lead; put in a very little japan gold size, and thin to workable consistency with turps; let it stand until tacky. It must be hard enough to prevent rubbing up or sweating. The method with the tip, gold knife and cushion requires considerable dexterity as well as practice to do good and rapid work. The tip, or lifter, is only a few camel hairs glued between two pieces of paste board, or other material. The knife is a long narrow flexible blade, and the cushion is made on a block, 6 by 8 inches, first covered with a thickness or two of woolen cloth, and finished by stretching a piece of chamois skin over it. Hold the gold book in the left hand, and turn back a leaf of the book, leaving the gold exposed on the next leaf; press the leaf of gold against the cushion and it will remain. Then straighten out wrinkles by a slight puff of the breath from above, cut the leaf into the required size with the gold-knife, and lift the leaf to its place with the tip. The tip will lift the gold better if occasionally drawn over the hair of your head.

Another way to prepare the leaf: Cut the book through at the binding with a sharp knife, which will leave all the leaves free and separate. Now take up the top paper or cover, which will leave the gold leaf on the book; lay the paper on a board and rub it over with a piece of wax, paraffine candle, or a piece of hard soap; either will do. Place the waxed side on to the gold, and smooth the paper down gently; repeat until you have as many leaves prepared as you need. Then, with good sharp shears cut them in such shape and size as will best cover your work, and not waste the gold. Lay the pieces on your board, gold side up. When ready, lay the pieces on the work, rub down with the fingers, or a ball of cotton, take off the paper and the gold will stay on the size. In this way the gold adheres quite firmly to the waxed paper, and the size must have a strong tack to take the gold off the paper. Experts lay the leaf directly from the book, and you had best learn to do it that way for general work, if you spoil half a dozen books while catching on to the knack of it. Try it this way: Now, here is a stripe half an inch wide, and the size is ready for the gold. Now hold the book flat in your left hand with your thumb on top, hold the top paper firm with your thumb. (If you let it slip, the leaf under it will be spoiled.) If the stripe is one-half inch wide, turn back enough of the paper to ex-pose three-fourths of an inch of the gold leaf, crease the turned back cover down with the fingers of the right hand, and hold it with the thumb on the back. Now cut the leaf with the finger-nail, first rubbing it dry on your pants; then turn the book carefully and quickly over on to the stripe, and press the gold down gently by pressing the book. Then turn down more of the paper, and repeat until that leaf is gone; then take another and so on. If the book gets too limber towards the last to handle well, have a square of cardboard to lay under the book next to the hand; you will find this is a help even with a full book. You will, perhaps, waste more gold in this way than by the transfer method, but you will more than make it up in time, if you become expert.

1st. Be sure of a good foundation.

2d. Have your gold size right, and study to know when the tackiness is just right. If your surface is not perfectly free from tackiness, pounce with a bag of gilder's whiting before putting on the size, to keep the gold from sticking outside of the size.

When you lay the leaf from the book and cut the leaf with your finger nail, turn the ball of the finger toward you and the nail towards the gold, and run the nail close to the edge of the turned paper; then, if the nail is not too long, the end of the finger will hold down the paper while the nail cuts the leaf.

To prepare paper for the transfer method I rub the paper on my hair, then lay it on the gold leaf, gently rub it with my finger tips, and the leaf adheres to the paper.

It can then be cut with shears in any desired shape to cover the work.

Some gold leaf is now packed in paper so prepared that the leaf will adhere to one side of it and can be taken up in that way.

Some gilders take up the leaf by wetting the paper on the back with turpentine to make the leaf adhere to the other side, when it can be cut to the required shape with shears. This is done instead of waxing the paper.

STIR YOUR PAINT.

It isn't always your material that makes a bad job, but it seems an easy matter to make even the best of paint the scapegoat for bad work. The heedless workman who primes a plastered wall without sweeping down the loose sand, or is careless about taking the sand and dust from the tops of casings and the floors, will, if he stops to examine, find some in the brush and some of it in his paint pot; and then, to cover up his carelessness, he can lay the blame on the paint. The careful painter will, when using heavy pigments, carry a paddle, and not neglect to use it. To prevent white lead and other heavy pigments from settling in the pot the paint must be well mixed, and kept mixed by stirring with a paddle as often and as much as may be necessary to *keep* the oil or other vehicle, and the pigment well incorporated. No one out a novice, or a careless painter will permit a sediment to accumulate in the bottom of his pot; no matter whether the pigment is coarse or fine; or whether the vehicle used is linseed oil, turpentine or benzine. The painter who goes to work without a stirring paddle in his pot will be liable to do uneven work, and find more or less sediment in the bottom of his paint pot at quitting time, because there is no white lead made which does not contain more or less particles sufficiently heavy to *commence settling* the minute the paddle stops, and go to the bottom of a pot of flating, as ordinarily mixed, inside of thirty minutes, and other particles of smaller size will follow later. If the pigment is mixed with oil the process of settling is slower, but no less sure to take place, and continue, if undisturbed, until clear oil stands on top of the pigment. Don't try to use your brush for a paddle; it isn't a good tool to stir paint from the bottom. Paint made of heavy pigment must be frequently stirred with a paddle to keep it of uniform consistency, but this operation is too often neglected. For instance, a man starts out with a full pot in the morning and neglects to stir his paint as he works, hence the heavier particles commence to settle and soon get below the dip of the brush, and by continual settling keep out of the reach of it until they reach the bottom. When the paint is nearly all out, and the sediment at the bottom don't work well, he refills his pot, leaving in the coarse pigment. At night the boss finds an inch or less of coarse paint in the bottom of the pot, and without further inquiry complains that the lead is sandy.

Another instance: The paint for a job stands mixed over night; the painters fill their pots from time to time during the day, but never stir the paint from the bottom, hence the last pot or two filled will have all the coarse pigment of the batch. There are cases, I admit (too many of them), where not only white lead, but dry colors and colors in oil, are too coarse to work well, but the best white lead and heavy colored pigments in oil or turpentine are liable to be called sandy unless frequently stirred by the painter.

TO MAKE CHERRY STAIN.

Take annotto, 4 ounces, and clear rain water, 3 quarts. Boil in a brass or copper kettle, new tin or galvanized iron will do, until the color of the annotto is imparted to the water; then add ⅛ ounce potash, and keep the mixture hot for 30 minutes; then, as soon as cool enough to handle, it is ready for use. Now, have the work free from dust, and spread on your stain with a brush or sponge and rub it well into the wood.

When the work is dry, rub lightly with fine sandpaper, because the water stain will raise the grain unless the wood has been filled.

You can suit the taste of the owner as to depth of color by repeating the operation, or by making the stain weaker or stronger, as the case may require.

VARNISH STAINS.

These often come very handy to the painter, not only in toning up new wood, but in renewing the freshness of old work.

MAHOGANY VARNISH STAIN.

Spirits 1 gallon, gum sandarac 1 pound, shellac ½ pound, venice turpentine 2 ounces, dragon's blood 4 ounces.

WALNUT VARNISH STAIN.

Shellac 1½ pounds, spirit 1 gallon, Bismarck brown 1 ounce, nigrosine ½ ounce. You can, by varying the proportions of the two colors, make the shade as you like it.

(Spirit in this connection means either wood or grain alcohol.)

MAHOGANY VARNISH STAIN.

Spirits 1 gallon, shellac 1½ pounds, Bismarck brown R ½ ounce, nigrosine 30 grains. More nigrosine will make the stain darker. If this is too thick to work well, thin with spirits.

TO MAKE NEW OAK LOOK OLD.

Sponge it with a strong hot solution of common soda in water. This will raise the grain, hence it will require cutting down with sandpaper.

DARK STAIN FOR OAK.

Make a solution of bi-chromate of potash, 1½ ounces to 2 quarts soft water. Lay on the solution with a good clean sponge and keep the wood wet with the solution until it is dark enough to please you. Then wash off the potash with clean soft water.

ANOTHER.

Apply with a brush, strong aqua ammonia until you get the desired shade.

RED SAUNDERS STAIN.

Fill a bottle ⅓ full of red saunders, then fill the bottle with either wood or grain alcohol. The more red saunders you put in, the stronger will be the stain; you can dilute it for the lighter shades. The longer it stands, the more color will be extracted. Always strain through muslin before using.

Red saunders makes a good cherry stain. When used on the bare wood it requires no binder, but when used over filled or oiled wood, put in one-fourth as much shellac varnish as you have stain, to act as a binder for it. If you want it to act as a filler as well as a stain, for pine or other close-grained wood, add 1½ pounds corn starch, to each gallon of the mixture of stain and shellac. Try a little and if it rubs up when dry, add more shellac.

You can mix red saunders stain with asphaltum varnish, to make black walnut and mahogany stains, using more or less of either to give the desired shade by using turpentine to make them mix. The asphaltum acts as a binder in place of the shellac.

The practical painter can get the shades he wants by experimenting on this line.

TO CHANGE THE COLOR OF WALNUT TO DARK MAHOGANY.

First give it a coat of very thin asphaltum varnish, then, when dry, give it a coat of red saunders and shellac.

You can mix the red saunders and asphaltum stain with any turpentine varnish, or with spirit varnish, if you use turpentine to make them mix.

Burnt umber and burnt sienna in oil or varnish make a walnut stain. Use but little of the pigments in proportion to the oil. Too much pigment gives the work a muddy color.

NATURAL WOOD FINISHING.

Clean up all soiled places on the wood. To be sure of a good job on open grained wood use a Bliss Rock Wood Filler. If you use a ready made filler, thin as per directions on the can. Whatever filler you use, put it on with a good brush. As soon as the filler begins to set, or show flat, commence to rub it into the grain with a pad made by gluing a piece of harness leather onto a block; always when practicable rub across the grain of the wood. For round work have a long piece of leather to draw back and forth around the work. Remember the main thing at this stage is to get as much of the filler as possible rubbed into the wood.

Another important point is to take off the surplus filler before it becomes too hard to wipe off, and another point is to wipe off the surplus filler and leave the pores of the wood level full. Hence, it is important that the filler does not dry too fast, that the painter puts on no more filler at a time than he can handle before it dries, and that in wiping off the surplus filler he works his rags across the grain. Some very open grained wood requires a second application of filler to make a good job, or at least to be looked over and touched up. The filler should have at least two days to dry. When dry go over it lightly with fine sandpaper to take off all particles of filler left on the surface.

Walnut, mahogany, chestnut, oak, ash and butternut may be classed as open grained woods, which need to be well filled with paste filler colored to match the color of the wood. When the filler is dry put on a coat or two of white shellac and rub down smooth with No. 1 sandpaper, and follow with two or more coats of hard oil or varnish, as you like; give each coat plenty of time to dry and rub each coat with curled hair or hair cloth, except the last coat. If you want an egg shell or half gloss, rub the last coat with pulverized pumice stone and raw linseed oil. If you want a dead finish rub down with pulverized pumice stone and water instead of oil. If you want a polish, first rub with the pumice stone and water; then with rotten stone and water, and polish with rotten stone and oil, or furniture polish and rotten stone. If you want a gloss finish, flow on the last coat and omit rubbing. Treat the close-grained woods as above stated, with the exception of the filler. The shellac also may be omitted, but it will take at least one more coat of hard oil or varnish for the job.

Cherry, sycamore, maple, birch, gumwood, redwood, cypress, pine, whitewood, poplar and hemlock are all close-grained woods, and need no paste filler. Pine especially should have a coat of shellac to keep back the pitch.

For an extra fine job of gloss finish, rub next to the last coat with pumice stone and water, flow on a coat of good varnish, and leave it in the gloss. In this case great care is required in cleaning the work to keep it from showing specks.

It stands the beginner in hand to be careful and not use his shellac too heavy to work well; shellac has good body and an apparently very thin coat will be a good heavy one.

To do a fine job the room and work must be clean, the clothing free from dust, and the work, brushes and varnish free from specks. If specks show on your gloss coat call a halt, and find where they come from.

Soft cotton rags are the best material for wiping off surplus filler.

A felt pad of convenient size to handle is the best for rubbing work. Get one at the furniture shop. For a cheap job omit the water rubbing, and rub with pumice stone and raw oil.

TO MAKE BLEACHED OR WHITE SHELLAC VARNISH.

Take powdered white shellac 1½ pounds, best grain alcohol 1 gallon. Add the gum to the alcohol, set it in a warm place and shake your jug or bottle occasionally. Don't put it in tin or iron; either of them will discolor it. You can hasten the process by setting your jug in a sand or water bath, and gently heating it; or set it by the stove, or in the sunshine.

To make the common orange shellac of commerce, dissolve 1½ pounds orange shellac in 1 gallon methylated spirit or grain alcohol. This will dry in ten or fifteen minutes, and makes a hard lustrous varnish when dry, and stands the weather better than most gum varnishes. It makes a turbid liquid of orange brown hue and dries rather a pale brown. For use on dark wood this is equal to the white shellac, if not superior.

TO COLOR PUTTY.

There is no use in trying to color common putty to match the color of natural wood. The whiting in it will not take clear tints. Use lead putty, which you can tint with raw sienna for pine, yellow ochre for oak, burnt umber and burnt sienna for walnut, and burnt sienna for mahogany. Better have the putty too light than too dark.

SPOTS ON PAINT.

Poor lumber and thin painting are often the cause of spots on paint, especially on two-coat work. On cross-grained and other extra-porous places more of the oil sinks into the wood than on the general surface, and the result is flat places in the paint, which fade sooner than the glossy paint; hence, the work looks spotted.

To provide against this kind of spotting use more care in priming and see that all extra-porous places are well filled with the prime coat, or touch them up before the second coat goes on. A little extra work with the brush when putting on the prime will save trouble.

Another cause may be traced to the practice of putting on a coarse dark priming coat, which will show through in places where the paint is thinnest.

Mildew, or fungus growth, is another cause. This sometimes comes from the use of too much japan, *poor or fat oil*, or when the paint dries tacky or soft.

Adulteration of linseed oil with mineral and other non-drying oils, has a tendency to make paint dry soft. Linseed oil, kept for a few days in an old sour tank or in an old rancid can in the paint shop, is liable to cause fermentation to take place, which may result in mildew in damp weather in shaded places.

When an oil can smells sour, or there is a deposit of foots at the bottom, it is unfit to keep oil in.

Another cause of spotting may be found in insufficient and improper brushing or spreading the paint; especially the priming, which requires as much care in putting on as any other coat on the job.

For instance, here is a job which shows "laps." Now, if this prime is right when it is put on single, it is wrong when it is put on double, because, where the laps are, the work has at least one more coat than the balance of the job, hence the paint is liable to fade spotted.

PORCELAIN FINISH.—CHINA GLOSS.—GLOSS FINISH.

All different names for about the same thing. To make a fine job: If the work is new, see that it is smooth, free from dust and stains. Then give it a coat of priming, put on thin, so as not to show brush marks, and rub down with No. 0 sandpaper. Next, get a good body with keg lead, mixed in turpentine and a very little linseed oil; put on thin coats, so as not to show brush marks; use a fitch brush, or at least a *fine* bristle chiseled brush. When dry, rub down with sandpaper and flow on a coat of thin white shellac. This is to keep back the oil in the lead coats, and prevent chemical action between the lead and zinc coats. Next, put on two or more coats of French zinc ground in damar varnish; enough at least, to get a clear white. Thin with turps and a little damar varnish, and put on thin enough to show no laps or brush marks.

Then put on a coat or two of French zinc ground in damar varnish, thinned with 1 part damar varnish and 2 parts turpentine. Next put on a coat of damar varnish mixed with a little zinc ground in damar, just enough to make the varnish white. Flow on a coat, and be careful that it does not run on your work. To avoid runs always commence at the top of a panel with a full brush and work down so as not to have a surplus in the lower corners of the panels; this applies to all parts of the work. It is quite a knack to put on a full coat of this varnish and zinc, and not have it run.

In all cases put on enough zinc coats to make a clear white before you put on the varnish. The small quantity of zinc is put in the varnish to take off the yellow tinge, and to keep it from turning yellow. Use lead putty. See recipes to make it on another page.

ANOTHER WAY.

Very hard and white, for parlors.—To prepare the wood for the finish, if it be pine, give one or two coats of the "Varnish—Transparent for wood," which prevents the pitch from oozing out, causing the finish to turn yellow; next, give the room, at least, four coats of pure zinc, which may be ground in only sufficient oil to enable it to grind properly; then mix to a proper consistency with turpentine or naphtha. Give each coat time to dry. When it is dry and hard, sandpaper it to a perfectly smooth surface, when it is ready to receive the finish, which consists of two coats of French zinc ground in, and thinned with damar varnish, until it works properly under the brush.

LEAD POISONING—HOW TO AVOID IT.

White lead may enter the human system in three ways, to-wit: Through the stomach, the lungs and the skin. In other words, it may be eaten, inhaled or absorbed, hence the stomach, lungs and skin should each be carefully guarded against it. To guard the stomach, through which you are in the most danger of taking in the poison, make it a rule to keep the mouth closed as much as possible when using white lead, and *especially when sandpapering.* Make it a rule to never eat or drink without first carefully cleansing your lips, and carefully removing the paint from your hands before eating. Tobacco chewers, who carry tobacco in their pockets, are in especial danger of lead poison, if working in paint, because the tobacco becomes more or less poisoned with lead from the fingers, if the painter is not careful to clean his hands before taking a chew. There is no great danger from inhaling white lead, except when sandpapering, or when dusting after sandpapering.

It is a pretty good thing to carefully guard the nose with a damp sponge while sandpapering, and to carefully free the nostrils from lead. There is no danger of poisoning by absorption through the skin, unless the painter is careless. When T see some men at work, I wonder how they can possibly escape lead poisoning. Their clothing glazed with oil paint, their hands daubed to the wrist by grasping the brush by the head, instead of by the handle; or by general carelessness in mixing and handling paints.

SYMPTOMS OF LEAD POISON.

Tired feeling, wakefulness at night, neuralgic pains, "shaky" hands, constipated bowels, bad taste in the mouth, and pain in the bowels, a blue edge on the gums, and a coated tongue. If you get the colic, see a doctor; for the other symptoms, get away from paint for a while if possible, and take the following: Iodide of potash, ½ oz.; syrup sarsaparilla, 8 oz. Dose:—Teaspoonful three or four times a day in half a cup of milk. Eat graham mush and drink milk.

TO FINISH FURNITURE AND OTHER WORK IN SIXTEENTH CENTURY OAK.

First fill the wood with any good filler. Fill it well, then take Vandyke brown 3 parts, and burnt sienna 1 part, and mix to a stiff paste with boiled oil and japan, and thin with turpentine, until you can brush it on the wood, and not have it look dauby or muddy. Give the work a light coat, and brush it out well and carefully. Too much pigment will make your work too dark. Wherever you want the light or worn spots to appear, wipe off the stain with a cloth, and with a badger blender carefully blend the stain into the edges of the worn or light spots. Don't stain too much at once, for fear your stain may set so you cannot wipe out and blend. When the stain is dry, sandpaper lightly with No. 0 paper. Finish with two coats rubbing varnish, or with hard oil finish. Polish with rotten stone and raw oil.

A SUPERIOR GLUE (WATERPROOF).

A very superior article may be made by dissolving 3 parts of india rubber in 30 parts of naphtha; heat and agitation will be required to effect the solution; when the rubber is completely dissolved, add 64 parts of finely powdered shellac, which must also be heated in the above mixture until all is dissolved. This mixture may be produced in sheets like glue by pouring it while hot upon plates of metal, where it will harden. When required for use, it may simply be heated in a pot till soft. Two pieces of wood or leather, joined together with this glue, can scarcely be sundered without a fracture of the parts.

A VALUABLE CEMENT.

We find the following recipe good: The compound of glycerin, oxide of lead, and red lead, for mending cast-iron that has been fractured with the happiest results. It takes some little time to dry, but turns almost as hard as stone, and is fire and waterproof. For mending cracks in stone or cast-iron ware, where iron filling cannot be had, we think it is invaluable. Take litharge and red lead, equal parts, mix thoroughly and make into a paste with concentrated glycerin to the consistency of soft putty, fill the crack and smear a thin layer on both sides of the casting so as to completely cover the fracture. This layer can be rubbed off, if necessary, when nearly dry, by an old knife or chisel.

LINSEED OIL AND IRON RUST.

The oleaginous principle of linseed oil is said to be in the nature of neutral salts called linolein, consisting of linoleic acid combined with a glycerine base. Linolein is said by some writers to constitute three-fourths of the volume of linseed oil, and that the drying properties of the oil reside in the acid principle of the linolein; that is, linoleic acid has the property of attracting and combining with oxygen to form the substance known as dry linseed oil. This acid is said to be a compound of several different acid principles, combined in definite proportions. Writers seem to disagree as to what the acids are, and in what respect they differ from the acid properties of the non-drying fixed oils, but that is a question which need not be discussed here. The glycerine base of linolein seems to be common to all fixed oils, and is set down as an oxide consisting of one equivalent of water and five of oxygen; hence the affinity between the linoleic acid and its glycerine base.

Linoleic acid, like other acids, has an affinity for alkalies and the ordinary metallic oxides. It unites with them, forming *neutral compounds*. This affinity is said to be electrical; the alkalies and oxides electro-positive, and the acid electro-negative. The greater the contrast in this respect, the stronger the affinity; hence, some acids separate others from their bases and form new salts by precipitation. As an instance:

Drop sulphuric acid into a solution of acetate of lead. It will displace the acetic acid, form sulphate of lead and precipitate, leaving the liberated acetic acid in solution. In linolein, this acid is so constituted that the affinity, or attraction between it and its glycerine base, is too feeble to resist and keep back the oxygen of the air; hence, when linseed oil is exposed to the air in a thin layer, oxygen unites with its linoleic acid, and this process continues until the oil becomes dry to the touch. Beyond this point the process is slower, because the oil is now less penetrable; but the process goes on until the layer of oil becomes hard and brittle, no matter with what pigment it may be mixed, although the pigment may for a time retard the action of the destroying elements.

Linseed oil dries too slowly for general use by the painter, hence various ways have been devised to hasten the drying process. If the foregoing theory is correct, the process which will cause the oil to dry to a good wearing body in the time desired, and leave it in the best condition to resist the action of the elements and the absorption of oxygen, is the best. I regard the lead oxides as the best dryers for this purpose—at least according to my experience. When we add an oxide to linseed oil as a dryer in the small quantity which experience has taught us is best to use, it is evident that it is not sufficient in itself to oxidize the whole of the oil to an appreciable extent. Writers differ as to the peculiar action of the oxides upon the oil, but I think it safe to say that the dryer sets up some chemical reaction which increases the affinity between the linolein and the oxygen of the atmosphere; at any rate, there is no dispute upon the point that linseed oil in drying absorbs a large per cent. of oxygen.

A knowledge of this unanimously conceded point led me to believe that a coat of pure linseed oil might make the best possible priming coat for iron work which had commenced to rust. Why? Because iron rust is an oxide of iron, having an excess of oxygen. Spread on rusty iron, it penetrates the rust, absorbs its excess of oxygen and dries with the remaining neutral oxide held fast in its body. This is my theory; whether correct or not, numerous tests have proved to me that a coat of linseed oil will stop the rusting of iron if applied under proper conditions. When rust is *thick or scaling* there is no safety short of taking it off. Iron rust is more or less hydrated; to free it from moisture, give it the flame of the gasoline paint burner.

WHITE ENAMEL (SELECTED).

First, the wood is primed with a composition consisting of three parts of turpentine and one part of oil, japan gold size being used as a dryer. On this drying thoroughly the work is rubbed down until perfectly smooth. Next are applied two or three coats of pure white lead mixed entirely flat; each coat is rubbed down, time being allowed for it to dry. Equal parts of lead and zinc are used for the next coat, and three-fourths zinc and one-fourth lead for the one succeeding. After this has become thoroughly hard it is rubbed down very smooth. A thin coat of color made of zinc and turpentine is now rubbed on; for the next coat the same flat color is used, with the addition of about one-half the quantity of good light coach varnish. For the last coat enough zinc is used in the varnish to make it white if the last coat of zinc is not white and solid before varnishing. If the work is to be gilded or striped the zinc must be left out of the last coat of varnish.

VARNISH TO IMITATE GROUND GLASS.

An expert has sent the following to the *British Journal of Photography*: To make a varnish to imitate ground glass, dissolve 90 grains sandrac and 20 grains of mastic in 2 ounces of washed methylated ether, and add, in small quantities, a sufficiency of benzine to make it dry with a suitable grain—too little making the varnish too transparent, and excess making it crapy. The quantity of benzine required depends upon its quality—from half an ounce to an ounce and a half, or even more; but the best results are got with a medium quality. It is important to use washed ether, free from spirit.

VARNISH FOR RUSTIC WORK.

One quart of boiled linseed oil and two ounces of asphaltum, to be boiled on a slow fire until the asphaltum is dissolved, being kept stirred to prevent its boiling over. This gives a fine dark color, is not sticky, and looks well for a year; or, first wash the article with soap and water, and when dry, on a sunny day do it over with common boiled linseed oil; leave that to dry a day or two, then varnish it over once or twice with hard varnish. If well done this will last for years and prevent annoyance from insects.

TO CLEAN VERY DIRTY BRASS.

Rub some bi-chromate of potassa fine, pour over it about twice its bulk of sulphuric acid, and mix this with an equal quantity of water. The dirtiest brass is cleaned in a trice. Wash right off in plenty of water, wipe it and rub perfectly dry, and polish with powdered rotten stone.

TO COUNTERFEIT TORTOISE SHELL VERY FINELY.

In order to do this well, your foundation or ground-work must be perfectly smooth and white, or nearly so, you then gild it with silver leaf with slow size, so as to have it perfectly smooth with no ragged edges, cleaning the loose leaf off. Then grind cologne earth very fine, and mix it with gum water, common size; and with this, you having added more gum water than it was ground with, spot or cloud the ground-work, having a fine shell to imitate; and when this is done, you will perceive several reds, lighter and darker, appear on the edges of the black, and many times lie in streaks on the transparent part of the shell. To imitate this finely, grind dragon's blood with gum water, and with a fine pencil draw those warm reds, flushing it in about the dark places more thickly, but fainter and fainter and thinner, with less color towards the lighter parts, so sweetening it that it may in a manner lose the red, being sunk in the silver or more transparent parts. When it is dry, give it a coat of varnish, let it stand for a few days, then rub it down with pumice stone and water. Then grind gamboge very fine, and mix with varnish, giving of this as many coats as will cause the silver to have a golden color, then finish with a clean coat of varnish.

PRICE LIST.

The prices of labor, and cost of material vary so much in different localities that it seems impossible to make a reliable price list for general work. The position, condition, and shape of different jobs all go towards making a general price list, an unreliable guide; also the quality of work demanded may make 50 per cent. difference in price. I have half a dozen printed price lists before me, and they generally agree to about the following prices for painting and glazing, to-wit:

	Per Yard.
1 coat on new work	8 to 10 cents
1 coat on old work	10 to 18 cents
2 coats on new work	18 to 20 cents
2 coats on old work	20 to 25 cents
3 coats on new work	25 to 28 cents
Brick walls, 2 coats	20 to 30 cents
Penciling	10 to 15 cents

PRIMING AND GLAZING SASH.

	Per Light.
10 × 14 and under	5 to 6 cents
12 × 16	7 to 9 cents
14 × 24	10 to 12 cents
18 × 24	15 to 18 cents
24 × 30	20 to 25 cents
30 × 40	35 to 50 cents

For old work where the old putty is in the sash, multiply the above figures by 3 or 4. When called out to the house to set a light or two charge for time and material. Most work of this kind is done at least 30 per cent. below the above prices.

I quote below a price list for sign painters, from a very complete report on painters' prices and measurements, generally, by one of the ablest of local associations of master painters and decorators:

JAPANNED TIN SIGNS.

	Gold.	*Plain.*
3 × 14 inches	$1.25	$ .75
6 × 8 inches	1.50	.75
8 × 10 inches	1.75	1.00
10 × 14 inches	2.50	1.50
11 × 17 inches	3.00	2.00
11 × 17 inches, 3 lines	3.50	2.55
14 × 20 inches	4.00	2.50
14 × 20 inches, 3 lines	4.50	3.00
18 × 24 inches	6.00	3.50
18 × 24 inches, 3 lines	7.00	4.00

Frames additional.

GLASS SIGNS ON WINDOWS AND DOORS.

In Silver or Gold	*Per Foot.*
Letters up to 6 inches in height	$ .75
Letters 6 to 10 inches in height	1.00
Letters 10 to 14 inches in height	1.50

Shaded, one color, 25 per cent. extra.

DRUM SIGNS.

	Gold.	*Plain.*
10 × 14 inches	$ 3.50	$ 2.50
11 × 17 inches	4.00	3.00

14 × 20 inches	5.00	3.50
18 × 24 inches	7.00	5.00
20 × 24 inches	8.50	6.50
24 × 30 inches	10.00	7.00
30 × 30 inches	12.00	8.50
30 × 48 inches	15.00	10.00

The above include moulding and urns and putting up.

Drilling holes in iron extra.

MUSLIN SIGNS.

	Per Foot.
Up to 1 foot high, black	8 cents
1 to 2 feet, black	10 cents
2 to 3 feet, black	12 cents

Colored one-half extra.

Muslin furnished.

Frames extra.

OIL CLOTH SIGNS.

	Per Foot.
Up to 1 foot	20 cents
1 to 2 feet	25 cents
2 to 3 feet	30 cents

Oil cloth furnished.

Frames extra.

BOARD SIGNS.

Including three coats of paint and lettering.

	Gold.	*Plain.*
6 inches × 4 feet	$ 4.00	$ 2.50
8 inches × 6 feet	5.00	3.50
10 inches × 8 feet	6.00	4.00
1 foot × 12 feet	7.50	5.00
1 foot × 15 feet	8.50	5.00
14 inches × 16 feet	8.50	5.00
14 inches × 18 feet	9.00	6.00
14 inches × 20 feet	9.50	6.00
16 inches × 16 feet	9.50	6.00
16 inches × 18 feet	10.00	6.00
18 inches × 18 feet	10.00	6.00
18 inches × 20 feet	12.00	7.00
18 inches × 24 feet	15.00	7.00
18 inches × 30 feet	18.00	8.00

Board extra.

Irons and putting up extra.

Shading, 25 per cent. additional, one color.

WALL SIGNS.

Two coats of paint and lettering. Extra coat, 1 cent per square foot additional.

2 × 16 feet	$ 6.00
2 × 20 feet	7.00
2 × 24 feet	8.00
2 × 30 feet	10.50

3 × 16 feet	8.00
3 × 20 feet	10.00
3 × 24 feet	12.00
3 × 30 feet	14.00
4 × 16 feet	9.00
4 × 20 feet	12.00
4 × 24 feet	13.00
4 × 30 feet	15.00
6 × 16 feet	12.00
6 × 20 feet	14.00
6 × 24 feet	16.00
6 × 30 feet	18.00
8 × 16 feet	14.00
8 × 20 feet	16.00
8 × 24 feet	18.00
8 × 30 feet	20.00
10 × 12 feet	10.00
10 × 16 feet	13.00
10 × 20 feet	16.00
10 × 24 feet	19.00
10 × 30 feet	22.00
12 × 16 feet	14.00
12 × 20 feet	18.00
12 × 24 feet	20.00
12 × 30 feet	25.00
14 × 20 feet	20.00

14 × 24 feet	24.00
14 × 30 feet	28.00
16 × 24 feet	26.00
16 × 30 feet	39.00
20 × 24 feet	30.00
20 × 30 feet	35.00
20 × 40 feet	40.00
21 × 30 feet	37.00
24 × 36 feet	42.00
24 × 40 feet	48.00
30 × 40 feet	60.00
30 × 50 feet	70.00
30 × 60 feet	80.00

SHOW CARDS.

1 sheet, 22 × 25	$ 1.50
1 sheet, 14 × 22	.75
1 sheet, 11 × 14	.50

The above prices are based upon white lead at 7 cents per pound and wages at 33½ cents an hour.

MIDSUMMER PAINTING.

All things considered, which is the best time of the year to do outside painting? Spring and fall, did you say? Well, yes. I know nearly all painters think so, and the people outside the trade are almost, if not quite, unanimous in holding the same opinion. But why? Do the winds of March, the frequent showers of April and May add very much to the pleasure and profit of doing outside work in spring? Do the soaking rains, which come along about the time of the vernal equinox and drive you off your job for a week or two and watersoak your unprimed work, add much pleasure to your recollections of spring painting? Do you remember anything about the clouds of midges and thousands of little moths which filled the air, ready and willing to decorate your paint with their little bodies on every still, warm mid-day in April and

May? Of course, we are speaking now of climatic conditions from our own standpoint, the great Northwest, which may also be true in the Middle and New England states. The mornings and evenings of spring and fall are apt to be cool—often frosty; then the oil stiffens and the paint rubs out hard and goes on slow, and we lose time and work harder. Practically, I favor midsummer for outside work, because the temperature is more uniformly warm and the paint spreads easily and evenly at any time of day, and as a rule the rains are less frequent and give a longer warning of their approach. The little black flies are not so plentiful in the hot days of summer as they are in spring and early fall. They are either dead or seek the shade of trees and grass. The dew is all gone in summer before seven o'clock a. m., and does not commence to fall until after quitting time. A carpet of grass and other vegetation covers a large portion of the ground in summer, holding down the dust. The winds are not usually so high and gusty in summer as they are in the spring and fall. In the warm days of summer your work is more apt to dry quickly, cleanly and evenly; and when you "knock off" from work at six p. m., and the sun is yet two hours above the horizon, you know that your last ground stretch will soon be out of the way of dust and rain. In the hot weather of summer the pores of the wood are all open, and the oil, which is then soft and thin, goes farther into the wood than in spring and fall, when the weather is cool. There are, it is true, some fine days in the fall for outside work, but the rainy season of the autumnal equinox and the frosty nights of the later months often retard your work and mar the finish of your job. One objection urged against summer painting is the flies, but really are the flies which injure paint any more numerous in midsummer than they are in spring and fall? It is true the festive house-fly is in his glory in the summer, but, as a rule, he is too smart to get stuck in outside paint. To get inside is his ambition, and the molasses-cup and sugar-bowl are his objective points. If the house-fly is an objection in the summer, it certainly is a greater one in the fall, for in September and early in October they are thicker, saucier and more familiar than at any other time of year; then they want not only to get at the sugar, but to get in and warm.

A correspondent asks: "Does the reader know from practical experiment that one season is better than another for applying outside paint?" I suppose the writer means the effect upon the wearing qualities of the paint and the permanency of the color. I have been experimenting for a practical solution of this question for my own satisfaction and guidance, and have come to the conclusion that paint put on the outside in the hot weather of summer will wear as well and hold its color as long as paint put on in the cooler days of spring and fall. I know the idea that paint dries too fast in hot weather is almost universal, but I think it grows largely from the fact that a quick-drying paint is not as good for outside as a slow dryer; but you must remember that there is a great difference between a quick-drying paint and drying a slow paint as quickly as the ingredients will admit of. Linseed oil dries or hardens by absorbing oxygen from the air, and that process goes on more rapidly in hot weather than in cool weather, because the air in hot weather is in a condition more freely to part with its oxygen, or because the oil is in a better condition to receive it, or both. In other words, a warm atmosphere hastens the process of absorption and a cool air retards it, but in either case the result is the same: the air gives up enough of its oxygen to solidify the oil. Now, the question arises, can any difference be discovered (chemical or otherwise) in the composition of the paint, whether dried in warm or cool air? From a business-point of view, I have long advocated summer as a good time to paint outside, and have usually succeeded in converting customers to my views upon the subject, and as a consequence have not often had a dull time in midsummer. We painters in the country know how unpleasant and unprofitable it is to have all the work of the year rushed upon us in the spring and fall, and I think if painters generally could convince themselves by practical experiment that, all things considered, summer time is the best season of the year to do outside work, and advocate the same to their customers, backed by argument and practical illustration, there would soon be less need of complaint about a dull season in midsummer.

TO REMOVE PAINT.

1. An expeditious way is by chemical process, using a solution of soda and quicklime in equal proportions. The soda is dissolved in water, the lime is then added, and the solution is applied with a brush to the old paint. A few moments are sufficient to remove the coats of paint, which may be washed off with hot water. The oldest paint may be removed by a paste of the soda and quicklime. The wood should be afterwards washed with vinegar or an acid solution before repainting, to remove all traces of alkali.

2. Wet the place with naphtha, repeating as often as required; but frequently one application will dissolve the paint. As soon as it is softened, rub the surface clean. Chloroform mixed with a small quantity of spirit ammonia, composed of strong ammoniac, has been employed very successfully to remove the stains of dry paint from wood, silk, and other substances.

3. To remove paint from floors.—Take one pound of American pearlash, three pounds of quickstone lime. Slake the lime in water, then add the pearlash, and make the whole amount about the consistency of paint. Lay the mixture over the whole body of the work which is required to be cleaned, with an old brush; let it remain for twelve or fourteen hours, when the paint can be easily scraped off.

TO SOFTEN PUTTY AND REMOVE OLD PAINT.

1. Take three pounds of quickstone lime; slake the lime in water, then add one pound of American pearlash; apply this to both sides of the glass and let it remain for twelve hours, when the putty will be softened, and the glass may be taken out without being broken. To destroy paint, apply it to the whole body which is required to be cleaned; use an old brush, as it will spoil a new one; let it remain about twelve or fourteen hours, and then the paint may be easily scraped off.

2. To remove paint from old doors, etc., and to soften putty in window frames, so that the glass may be taken out without breakage or cutting, take one pound of pearlash and three pounds of quicklime, slake the lime in water and then add the pearlash, and make the whole about the consistency of paint. Apply to both sides of the glass and let it remain for twelve hours, when the putty will be so softened that the glass may be taken out of the frame without being cut, and with the greatest facility. To destroy paint, lay the above over the whole body of the work which is required to be cleaned, using an old brush, as it will spoil a new one. Let it remain for twelve or fourteen hours, when the paint can be easily scraped off.

3. Paint stains on glass.—American potash, 3 parts; unslaked lime, 1. Lay this on with a stick, letting it remain for some time, and it will remove either tar or paint.

TREATMENT OF DAMP WALLS.

There are two classes of damp walls, first where the water comes in from the outside from defective roofs, bad gutters, defective pipes, and where it comes through the walls from the ground, as in basements. In the other class we may include walls which are dampened by condensation of moisture, in places shut off from the general artificial temperature of the room, behind stationary furniture. Such walls may dry out during hot weather, or they may be kept damp by a growth of mold or fungus.

When water comes in from the outside, it is impossible to keep paint or paper on the wall in good shape. Look around for the places where the water comes in, point it out to the owner, and if he fails to stop the leak have it *understood* that the work is done at *his risk*; or, what is better, refuse to do the work; because, when a job comes off, or turns out badly, you will take the blame generally, no matter whether it is your fault or not. A job may be made to last awhile by a waterproof coating, or by sheathing with thin lumber, but it is only a question of time when the lining material will become water-soaked and spoil the paint or paper, to your discredit. I have usually been *too busy* to take jobs of this kind. If the water can be cut off, the next thing is to dry the wall, which you can do at the surface only by setting a stove near it, or with the flame of a paint burner; then, after all your trouble, the water, which remains in the wall, if of brick or stone, may find its way to the surface, and destroy your work. Sheet lead cemented to the wall will answer a good purpose for a time, but the dampness will finally destroy the cement and let the metal loose.

Battening out for lath and plaster is the best for basement or damp stone walls, but that is the plasterer's work, and is rarely ever done except in private residences.

Battening and canvasing is next best; nail your battens up and down 18 inches apart. Have the canvas stitched in sheets the right size to cover the large blank spaces of the wall. Then stretch and tack it on the battens, and give it a coat of glue and alum size.

When dampness is caused by condensation the best remedy is to remove the cause and dry the wall.

TO PAPER ON A BOARD PARTITION.

When paper is pasted on boards, it must crack, when the lumber shrinks. If you paste cloth over the cracks, it must crack, if the cracks open further than the cloth will stretch. When you tack cloth on a partition and size it, if the size goes through the cloth and sticks it fast to the boards, it will be likely to crack when the lumber shrinks. For a good job I would advise that you first cover the partition *with paper tacked on*, then when you size the cloth, it will stick to the paper, and not to the boards. I have met with uniform success in this way; because the boards are left free to shrink and swell without breaking the cloth or paper. I like to sew the cloth together with a running seam in pieces large enough to cover all broad spaces, turn the smooth side out, stretch well, and fasten the edges only; drive the tacks an inch from the edges of the cloth, so that you can fasten them down smoothly with paste. When a man has been unwise enough to put a board partition across one end of an otherwise fine room, and is willing to pay for his folly: first, take measurements of the blank spaces, and sew together some fairly strong *unbleached* muslin, stretch on frames, and give it a coat of glue and alum size, and whiting; when dry, carefully fit each piece in its place and tack it an inch from the edges and fasten the edges down smooth with strong flour paste. Tack only at the edges, and if you are careful to butt edge the different pieces over the doors, etc., you can make a nice smooth job in this way. By using this method the paste will not stick the cloth to the wall. Use tinned tacks to prevent rust.

SANDPAPERING.

This is a job none of us like very well, but since it must be done, it is worth while to be able to do it to the best advantage. The first thing to look for is good paper. To test the strength of the sand, rub two pieces together, and if the sand don't fly off, it is good in that respect; next see if the paper is tough and will not tear easily. Chalk the back of your paper before you double it and it will not slip. Don't lose time using old, worn-out paper. New paper will, of course, cut faster than old paper, and the difference in the time gained by using sharp paper will pay for the new paper twice over. Using old dull paper is like trying to save money by using an old stub brush. Better use up fifty cents' worth of paper than to fool away dollars' worth of time trying to save money by using old paper.

If you have old, hard paint to cut down, which dry sandpaper will not touch, keep the work wet with benzine, and you will be surprised to see how fast the sandpaper will cut the paint. To put on benzine use a small spring-bottomed can, such as is used for oiling machinery. You can use any grade of sandpaper, and it will not soak up or gum. No. 1 paper is the best for this purpose. A good deal of time may be lost where scrapers could be used to much better advantage. A broad, flat scraper to shove endwise is always in order, and a few narrow ones with various shaped ends to fit in headings, moldings, etc., are a great help.

A STENCILED BORDER.

This makes a nice finish for a painted or kalsomined room. To make it look at its best, paint a stripe as wide as your stencil in a pleasant contrast to the paint on the room and put the stencil on that in soft harmonizing colors.

REPAINTING SCALED WORK.

To repaint a job which has commenced to scale, without taking off all the old paint, is very uncertain work, but if you have to try it, have it understood in writing, or before witnesses, that it is done at the owner's risk. First scrape off the loose paint, then go over the job with raw oil; put it on freely and let it stand until dry; then scrape off all the paint loosened by the oil, and coat up with strictly pure white lead and oil. Avoid zinc, and mixtures of zinc, and barytes, on jobs of this kind; because they are more or less liable to crack, and pull off more of the old paint. White lead and oil lightly tinted will hold it if anything will. Use raw oil and a little good japan.

TO MIX WATER COLORS.

Light weight colors which will not mix well with water may be easily mixed to a stiff paste with molasses or sirup, then mix in glue size for a binder and thin with water.

TO SIZE MUSLIN FOR LETTERING.

Use a thin size of white glue in water, or a thin starch paste. For a sign to stand weather, dissolve white wax in turpentine by heat. Melt the wax in a kettle, then take it outside and by degrees add sufficient spirits of turpentine and make a thin size.

One ounce of wax to the quart of turps is about right. Put it on warm with a brush.

ANOTHER FOR WHITE WORK.

Slake a little good, fresh lime in hot water and mix a size with skim milk. Then strain through cheese cloth. This size is, when dry, insoluble in water and will hold lettering as long as the cloth lasts. May be tinted.

A B C D E F
G H I J K L M
N O P Q R S T
U V W X Y Z &,.
a b c d e f g h
i j k l m n o
p q r s t u v
w x y z 1 2 3
4 5 6 7 8 9 0

No. 4. OLD STYLE EXTENDED.

A B C D E F
G H I J K L M
N O P Q R S T
U V W X Y Z & , .
a b c d e f g h
i j k l m n o
p q r s t u v
w x y z 1 2 3
4 5 6 7 8 9 0

TEST OF THE PURITY OF WHITE LEAD.

The following is an infallible and simple commercial test of the purity of white lead:

Take a piece of firm, close-grained charcoal, and near one end of it scoop out a cavity about half an inch in diameter and a quarter of an inch in depth. Place in the cavity a sample of the lead to be tested, about the size of a small pea, and apply to it continuously the *blue* or *hottest* part of the flame of the blow-pipe; if the sample be strictly pure it will, in a very short time, say two minutes, be reduced to metallic lead, leaving no residue; but if it be adulterated, even to the extent of 10 per cent. only, with oxide of zinc, sulphate of baryta, whiting or any other carbonate of lime (which substances are the principal adulterations used), or if it be composed entirely of these materials, as is sometimes the case with cheap lead (so-called), it cannot be reduced, but will remain on the charcoal an infuscatible mass.

A blow-pipe can be obtained from any jeweler at small cost. An alcohol lamp, star candle, or a lard oil lamp furnishes the best flame for use of the blow-pipe. This test is very simple and anyone can very soon learn to make it with ease and skill.

POLISH TO RENOVATE VARNISHED WORK.

One quart good vinegar, 2 ounces butter of antimony, 2 ounces alcohol, 1 quart oil. Shake before using.

BRONZES—COLORS.

White,	Silver,	Flesh,
Light Gold,	Dark Gold,	Rich Gold,
Lemon,	Orange,	Fire,
Copper,	Carmine,	Crimson,
Lilac,	Violet,	Brown,

Light and Dark Greens.

BLACK VARNISH FOR IRON.

Asphaltum, 2 pounds.

Boiled linseed oil, 1 pint.

Spirits turpentine, 2 quarts.

Melt the asphaltum with the oil in an iron kettle. Stir well before removing from the fire. When partly cool add the turpentine and a little good japan.

TO FREE BENZINE FROM ITS OFFENSIVE ODOR.

To deodorize benzine, add 3 ounces quicklime to the gallon of benzine; shake well. Let the lime settle and pour off and filter the benzine.

PAINT TO PREVENT WOOD EXPOSED TO THE GROUND FROM ROTTING.

Take of linseed oil, 4 parts; whiting, 40 parts; rosin, 50 parts; clean sand, 300 parts; heat together in a kettle until the rosin melts; then add 2 parts sulphate of copper; the mass to be well stirred, and thinned to workable consistency with linseed oil.

RECIPES FOR BLACKBOARD SLATING.

Dissolve 1 pound shellac in 1 gallon 95 per cent. alcohol; then add ½ pound best powdered ivory black, 5 ounces finest emery flour, 2 ounces ultramarine blue; mix well and keep air tight. When using stir frequently. If thick enough to show brush marks, add more alcohol; work quick with a fine brush.

TO MAKE A BLACKBOARD ON COMMON PLASTER.

Stop all cracks and holes with plaster paris mixed in glue size. When dry sandpaper until all is smooth; then paper the wall with white blank wall paper, butt the edges, put on with strong paste, and be careful to rub out all blisters. When dry prime with oil paint, then sandpaper with fine paper, and put on two coats of above slating. This makes an excellent blackboard. Boards which I made in this way twenty years ago are in good shape yet, and will last for years to come with an occasional repainting.

CHEAP SLATING, BUT GOOD.

Mix lamp black, 4 parts; ultramarine blue, 1 part, by weight, in turpentine, with sufficient good japan and a very little oil to bind it, then add one part by weight of *fine pumice stone*. Have it thin enough to flow on and not leave brush marks.

WATERPROOF OIL RUBBER PAINT FOR CLOTH.

Melt 2½ pounds of india rubber in ½ gallon of boiled oil by boiling. If too thick, add more oil; if too thin, add more rubber, and a little japan to dry it. Apply warm.

TO CLEAN PAINT.

Have some whiting on a plate, then dip a piece of flannel in warm, soft water and squeeze nearly dry, then take up some of the whiting by dipping the flannel in it, and rub the paint until it looks clean, then rub dry with a soft cloth or chamois skin.

GOOD QUICK STAIN FOR A BRICK CHIMNEY.

For red stain, take Venetian red, 2 parts; yellow ochre, 1 part—both dry—and mix with skim milk. For yellow stain, use water-lime, tinted with yellow ochre. Mix as above.

Skim milk when mixed with common quicklime, Portland cement, or Venetian red, is converted into an insoluble binder, which renders the mixture waterproof, so that it will not wash off when wet; neither will it rub up when dry. Other pigments can be added, by way of coloring, up to 25 per cent., without affecting the insolubility of the paint.

For a brick wall, which has not been rubbed or painted, Venetian red toned down with yellow ochre, beats any glue and acid mixture for durability.

TO CLEAN DOOR PLATES.

Put on with a rag a weak solution of ammonia in water, and rub to dryness.

TO CLEAN VARNISHED PAINT.

In a gallon of water, boil a pound of wheat bran, and wash the varnish with the water.

SLOWING THE DRYING OF PAINT.

In wall painting or otherwise, especially in hot weather, if the paint dries so fast as to show laps in spite of your best efforts with the brush, the addition of a little cotton seed oil will make the paint dry slower without hurting the gloss; or if you are using flat color, and it dries too fast, a little cotton seed oil will make it dry slower, and not make a gloss. You can, by a little experiment, determine how much of cotton seed oil to use in each case.

FINE BRONZE FOR METALS.

Red aniline (fuchsine), 20 parts; purple aniline, 10 parts; 95 per cent. alcohol, 200 parts; acid benzoic, 10 parts. Dissolve the colors in the spirit in a porcelain vessel in a water or sand bath; add the acid and boil until the mixture changes from a greenish color to a beautiful bronze color. Lay it on the bright metal with a brush.

REPAINTING BLISTERED DOORS.

When the paint commences to blister or scale on a door, it is very liable to keep on blistering and scaling from time to time, as long as any of the old paint is left on the door, no matter how carefully it may be repainted, because in most cases whatever caused the paint to scale off in spots, weakened the entire coat of paint on the door, making it liable to raise up, or come off in other places, whenever exposed to any extra strain, such as sun heat, or the drying of new coats of paint or varnish over it; hence, to have a sure thing on painting a scaled or blistered door, take off all the old paint. Put on a thin prime of pure white lead and linseed oil; use the priming sparingly *and rub it out thin*; let the prime dry and coat up with lead and oil paint, mixed with good body; put in a little turps and spread the *paint out thin*, so it will dry solid; rub each coat in the same way; give each coat time to dry solid. For work to be varnished, prime as above, and coat up flat. I think blistering is often caused by flowing on too much paint having too much oil in it, in proportion to the pigment, hence it does not dry solid, the oil is softened and expanded by heat, and the coating, which is more of an oil skin than a body of paint, lets go its hold on the wood and puffs out in a blister to make room for the softened and expanding oil skin. If painters will mix their paint with good body, and use more elbow grease in rubbing it out, they will have less trouble with blisters.

FIREPROOF PAINT FOR ROOFS, ETC.

A recipe published thirty years ago in the Maine *Farmer*:

Slake stone lime by putting it into a tub to be covered to keep in the steam. When slacked pass the powder through a fine sieve, and to each 6 quarts of it add 1 quart rock salt, and water, 1 gallon; then boil and skim clean. To each five gallons of this add pulverized alum, 1 pound; pulverized copperas, ½ pound; then slowly add powdered potash, ¾ pound; then add hardwood ashes sifted, 4 pounds; now add any color and apply with a brush. This paint stops small leaks in roofs, prevents moss, is incombustible, and renders brick waterproof. It is durable as stone.

VARNISH FOR IRON.

Genuine asphaltum 8 pounds, melt in an iron kettle, slowly adding boiled linseed oil, 5 gallons; litharge, 1 pound, and sulphate of zinc, ½ pound; continue to boil three hours, then add dark gum amber, 1½ pounds, and boil two hours longer. When cool thin with turpentine to good working consistency.

BLACK VARNISH FOR IRON.

Genuine asphaltum (not coal tar imitation), 1 pound; lamp black, ¼ pound; rosin, ½ pound; spirits turpentine, 1 quart. Dissolve the asphaltum and rosin in the turpentine, then rub up the lamp black with linseed oil, only sufficient to form a paste, and mix with the others.

TO MIX DRY LAMP BLACK.

First cut it up in benzine or turpentine to a thick paste, stir well and add linseed oil; if the black is to be used as an oil paint, a little at first, stir well and you may add more. In this way you will have no trouble in mixing it with other paint, if you do it when the paint is rather stiff.

TO CLEAN BRASS.

One-half ounce oxalic acid, 3 ounces rotten stone, ¼ ounce gum arabic, each in powder; made into a paste with sweet oil. Use sparingly and rub dry with flannel.

DIPPING PAINT.

Grind dry colors in japan and turps, with only enough japan to bind the pigment. When dry varnish, use any pigment you like, or use bolted whiting and color as you like.

TO MAKE WAX FINISH FOR FLOORS.

Take 2 ounces pearlash and 2 pounds white wax. Slice the wax thin, and boil it with the pearlash in 2 quarts of water; stir until the wax is melted and unites with the water.

Put on the finish with a brush, and polish with cloth or plush.

This finish will be good only for light service.

SPIRIT VARNISHES.

There are numerous recipes which might be given here for making the fine elastic varnishes, but it would not be practicable for the painter to make them, even if he had the requisite skill and experience, but with spirit varnishes it is very different, and the painter can make them by a formula as well as an expert can. (For formulas for white and orange shellac varnish see article on wood finishing.) For inside work, where the family is living at the time the work is being done, the alcohol varnish is preferable. First, because it dries very quickly, and second, because it is free from sickening or disagreeable odors.

Below are several recipes for making varnishes, which dry hard and lustrous. The spirit used is wood or grain alcohol; in either case, the spirit should be 95 per cent. proof.

BROWN HARD SPIRIT VARNISH (SELECTED).

1. Sandarac, 1 pound; shellac, ½ pound; gum elemi, 4 ounces; Venice turpentine, 4 ounces; spirit, 1 gallon.

2. Gum sandarac, 1½ pounds; shellac, 1 pound; spirit, 1 gallon. After the gums are dissolved, put in rosin turpentine varnish, 1 pint. This makes a good varnish, not as quick drying as pure spirit varnishes.

A brown varnish may be made by mixing shellac, 1½ pounds; pale rosin, 1½ pounds; spirit, 2 gallons.

WHITE HARD VARNISH.

1. Sandarac, 2½ pounds; gum thus, 1 pound; spirit, 1 gallon.

2. Mastic, ½ pound; sandarac, 2 pounds; elemi gum, 4 ounces; spirit, 1 gallon.

3. Mastic, ½ pound; sandarac, 1 pound; turps, 2 ounces; spirit, 1 gallon.

These are all prepared by mixing and setting in a warm place until the gums are dissolved, then they are ready for use. Shake occasionally. For fine work strain carefully.

PURE WHITE VARNISHES.

1. Pale manila copal, 8 ounces; gum camphor, 1 ounce; mastic, 2 ounces; venice turpentine, 1 ounce; spirit, 1 quart.

2. Sandarac, 8 ounces; mastic, 2 ounces; Canada balsam, 4 ounces; spirit, 1 quart.

3. Sandarac, 8 ounces; damar, 4 ounces; gum thus, 8 ounces; manila copal, 8 ounces; elemi, 8 ounces; spirit, ½ gallon. This is a good pale article.

4. Gum thus, 8 ounces; gum benzoin, 4 ounces; manila elemi, 4 ounces; spirit, 1 quart.

VARNISH PAINTS.

These are made by mixing opaque pigments with almost any varnish, using sufficient turps to make them spread well.

GOLD VARNISH.

Shellac, 8 ounces; sandarac, 8 ounces; mastic, 8 ounces; gamboge, 2 ounces; dragon's blood, 1 ounce; turmeric, 4 ounces; spirit, 1 gallon.

FURNITURE VARNISH.

Shellac, 1¾ pounds; sandarac, 4 ounces; mastic, 4 ounces; spirit, 1 gallon.

DAMAR VARNISH.

Damar, 1 ounce; sandarac, 5 ounces, mastic, 1 ounce; turps, 20 ounces. Digest at gentle heat until dissolved. If necessary add more turps to bring down to the proper consistency.

LACQUERS FOR BRASS AND TIN.

Pale gold lacquer.—Spirit, 1 gallon; orange shellac, 1 ounce; gamboge, ½ ounce.

Deep gold.—Orange shellac, 10 ounces; turmeric, 4 ounces; gamboge, 4 ounces; dragon's blood, ½ ounce; spirit, ¾ gallon.

Brass lacquer.—Shellac, 14 ounces; turmeric, 4 ounces; annotto, 1 ounce; saffron, ½ ounce; spirit, 1 gallon.

LEATHER VARNISH (BLACK).

Shellac, 12 ounces; gum thus, 5 ounces; sandarac, 2 ounces; lamp black, 1 ounce; turpentine, 4 ounces; spirit, ¾ gallon.

Mix the ingredients, and give them time to dissolve in the spirit in a warm place. A shake-up now and then will quicken the process.

PAPER HANGER'S OUTFIT.

Bib overalls, large pocket in front, side pockets for rule and shears, long trimming shears, shorter wet shears, straightedge, paste board, plumb bob, rule, paper brush, paste pail, size kettle, step-ladders and rollers, some sandpaper, soft cloths and long blotting paper to use under your roller on seams, when needed, and a plank for scaffold, when papering ceilings. For common sized rooms two step-ladders are good in the place of trestles to hold up the plank. For butt edging I can recommend James Marks' paper cutters. See description on another page.

PAPER HANGERS' PASTE.

Beat up four pounds of sifted wheat flour in cold water sufficient to make a stiff batter; beat out all the lumps, then add enough cold water to make it like pudding batter. Then pour in a little hot water and stir, then pour in hot water fast, and stir until the paste swells and thickens, and turns darker. It is then cooked. To keep the paste from "going back" and staining the paper, add about two ounces of powdered or well pounded alum to the boiling water which you pour on the batter. This will make three-quarters of a common wooden pail full of paste. It will do better and go further if you let it cool before using. Turn a little cold water on the top to prevent it skinning over while you wait for it to cool. When ready to use it, thin with cold water, until it works easily under the brush, and according to the wall. A very rough porous wall needs a stout paste and plenty of it, while a hard, smooth wall should have the paste thinned and less of it. I have known paper to crack and fall off from a smooth wall, because too much or too thick paste was put on. Just enough to cement the paper to such a wall is best; a body of paste between the paper and plaster will decay and peel off, and take the paper with it. The other extreme must be avoided also. Some hangers prepare this paste without the alum.

If hanging paper on a glossy painted surface, leave out the alum and add one-half pint of nice clear sirup to each gallon of paste.

TO MAKE A PASTE FOR PAPERING OVER PAINTED OR VARNISHED WALLS.

In a kettle mix some flour in water in the same way as in the above formula, but make the batter thinner. To each gallon of the batter add one ounce of powdered resin. Set the kettle on a moderate fire, and keep stirring it until it boils and thickens, and the resin is melted into the paste. When cool, thin down with a weak solution of gum arabic.

LIQUID GLUE.

Fine glue dissolved in alcohol makes a nice binder for fine water colors.

TO CRYSTALIZE GLASS.

Lay the glass flat and flow heavy alum water over it. Let it dry.

SIZE FOR WALLS BEFORE PAPERING OR KALSOMINING.

One pound good white glue, 1 pound good bar soap, 2 pounds pulverized alum. Dissolve each separately in one quart boiling water, having first soaked the glue. Mix the glue and soap water, and then slowly add the alum water, stirring all the time. Add cold water to make one gallon.

STAIN OAK WOOD.

Wash with a solution of bi-chromate of potash and acid water. One ounce to a quart of water.

SIZING WALLS.

"Anybody can do it!" Yes, but it takes an expert to do it right. It is not a difficult matter to make paper stick to whitewash, but the whitewash splits as far in as the paste goes, and a part of it invariably sticks to the paper when it comes off and a part of it is left on the wall. As a rule, if you size whitewash with flour paste and let it stand a few days it will crack and roll up. Now, pure glue size does not have this effect upon whitewash, but, on the contrary, it not only acts as a binder, but as an intervening coat between the paste and the whitewash. In other words, the glue size will stick the whitewash fast without causing it to crack, and the paste will adhere to the glue size without bad effects upon either. Now, in order to bind the whitewash, the glue should penetrate as far as possible. Hence, the size should be put on warm, and the room should be warm, otherwise the glue will get cold and stiff like jelly before it has time to penetrate; hence it will remain on the surface instead of going into whitewash as a binder. The idea is to get all you can into the wall and leave as little as possible on the outside. Another thing to look after is the quality of the glue. Very much of the white glue found on the market is not genuine glue. Some of it is adulterated with starch and white clay, some of it is not glue at all. A glue which will dissolve in cold water is not good glue, or if it melts readily in hot water without being soaked an hour or two in cold water, it is not first-class. If it has a dead white look it is not good. Good glue should be glossy and semi-transparent, and should soften and swell in cold water, but not dissolve in it. When put into hot water without being first soaked in cold water, it should not dissolve at once, but form into a lump and resist the action of the hot water for some time.

HOW TO APPLY WHITE ENAMELED LETTERS TO GLASS.

An extract from a circular issued by the manufacturers of these letters:

Having thoroughly cleaned the window and freed it from grease, draw with white marking chalk on front of it the plan or arrangement of outline it is intended to adopt—straight or curved, as the case may be. A rule is used for marking the straight lines and a piece of twine for the curved lines. Now divide these guide lines up into as many spaces as there are letters to go on, carefully proportioning them. Then apply the cement to the back of the letters with a knife, laying on equally around both the inside edges. Place the letter upon the window in the space marked for it and work it up and down, back and forth, pressing against the glass, so as to expel the air and secure a good adhesion, and taking care to press equally on top and bottom of the letter, as otherwise there is a likelihood of breaking. It is advisable, in cementing larger sized letters than six inches, to leave the letters lay for an hour after placing the cement around the edges, and then to give another coat of cement and attach the letters immediately. The object is to prevent all the cement from working inside the concave parts of the letters. In affixing larger and heavy letters, small pieces of beeswax (or, in summer, sealing wax) should be employed to keep them in position until the cement sets. As soon as the letters are attached to the glass take a small stick of wood, sharpen it on the end and clean away all superfluous cement, keeping the end of the stick constantly wet. Particular care should be taken to leave no openings between the letters and the glass (especially around the top edges) which would allow water to get in between.

If wax has been used, remove it after a few days and clean with a rag. The sign is then complete for long service. The above method will answer equally well on any smooth surface such as stone, iron, marble, wood.

To make the cement, mix two parts of white lead ground in oil with three parts of dry white lead, and thin it down to the consistency of soft putty with some good furniture or copal varnish. Then take small parts of it and grind them on a stone or glass plate in the manner of painters grinding color with a bowl or palette knife. This is to be continued until the cement is entirely smooth and cornless, and then it is ready for use.

To remove enameled letters, the most convenient way is to scratch away around the edges all the cement you can from under the letters. Use for this purpose a very thin knife or a piece of thin sheet steel. You will soon reach the soft part of the cement; then cut away with a sawing motion and twist them off. Do not attempt to pry the letters off, or they may break. If the cement should be very hard, say after a number of years, use a little kerosene oil, which is applied on the top edges of the letters, so as to work in and soften the cement.

WALL SIZING FOR KALSOMINING.

There are many things about wall sizing, which depend largely upon good judgment for success, because the treatment must be varied according to the condition of the wall or ceiling. A good size is made of good white glue, ½ pound; alum, 1 pound.

Dissolve the glue in the usual way; that is, soak it in cold water until soft, then pour off the cold water and pour on the hot water; and stir until the glue is dissolved.

Dissolve the alum in hot water.

Then stir the glue, and put in the alum water. Thin the mixture with water to the right consistency to work well.

If one coat is not sufficient, give it two; or if there are porous places in the wall, touch them up.

In many cases a simple glue size is sufficient, but if you use the glue and alum size as above directed, you will be pretty sure of a good foundation for kalsomine.

One of the most difficult things to overcome in preparing ceilings for kalsomine is the water stain, which is liable to be invisible until developed by a coat of kalsomine. If you find water stains on a ceiling and suspect that there may be others which do not show, go over the ceiling with a thin wash of whiting mixed in clear water, which when dry will develop all hidden stains. To kill a bad stain, first put on a coat of oil, japan and turps, equal parts; second, put on a coat of good heavy shellac; third, give the spots a coat of flat lead. This treatment is for dark stains; for light stains a coat or two of shellac will stop the stain. It is best to put a coat of keg lead thinned with turps over the shellac, because kalsomine is liable to scale off from shellac.

On cheap work, if the stain is not too dark, it may be kept back by pasting a piece of paper over it. If the wall has been kalsomined it is always in order to wash off the old kalsomine. If the work has been whitewashed, either take it off or first give it a wash of strong vinegar, then a glue size, which, if put on thin and plentifully while warm in a warm room, is about the best size I know of for whitewash. I have often used it successfully when it was not practicable on account of the weakness of the ceiling or other cause to take off the old whitewash. Two thin coats of good glue size on firm whitewash makes as fair a foundation for kalsomine as can be made on old whitewash.

When it will not pay you to wash off the old kalsomine, a coat or two of the wall sizing described above will make a good foundation.

SIGN PAINTING.

To the beginner I will say: Learn the letters; get a variety of alphabets in your head; the more you have the better you will be prepared to do a pleasing variety of sign writing. A variety of letters arranged in alphabets are given in the following pages as a convenient means of reference for the painter who may desire to refresh his memory, as to the form of any letter represented, or to make a study of them with a view of acquiring a knowledge of the formation of letters generally.

A B C D E F G

H I J K L M N

O P Q R S T U

V W X Y Z & .

No. 1. GOTHIC CONDENSED
A B C D E F G
H I J K L M N
O P Q R S T U
V W X Y Z & .

a b c d e f g h
i j k l m n o p q
r s t u v w x y z
1 2 3 4 5 6
7 8 9 0 , .

No. 1. GOTHIC CONDENSED—continued.
a b c d e f g h
i j k l m n o p q
r s t u v w x y z
1 2 3 4 5 6
7 8 9 0 , .

A B C D E F G H
I J K L M N O P
Q R S T U V W X
Y Z &
1 2 3 4 5 6 7 8 9 0
a b c d e f g h i j
k l m n o p q r s
t u v w x y z , .

No. 2. BLANCHARD.
A B C D E F G H
I J K L M N O P
Q R S T U V W X
Y Z &

1 2 3 4 5 6 7 8 9 0
a b c d e f g h i j
k l m n o p q r s
t u v w x y z , .

A B C D E F G H
I J K L M N O P Q
R S T U V W X Y Z &
1 2 3 4 5 6 7 8 9 0 , .

No. 3. ALASKAN.
A B C D E F G H

I J K L M N O P Q
R S T U V W X Y Z &
1 2 3 4 5 6 7 8 9 0 , .

A B C D E F
G H I J K L M
N O P Q R S T
U V W X Y Z & , .
a b c d e f g h
i j k l m n o
p q r s t u v
w x y z 1 2 3
4 5 6 7 8 9 0

No. 4. OLD STYLE EXTENDED.
A B C D E F
G H I J K L M
N O P Q R S T
U V W X Y Z & , .
a b c d e f g h
i j k l m n o
p q r s t u v
w x y z 1 2 3
4 5 6 7 8 9 0

A B C D E
F G H I J
K L M N
O P Q R S
T U V W
X Y Z & , .
1 2 3 4 5
6 7 8 9 0

No. 5. LINING GOTHIC.
A B C D E
F G H I J
K L M N
O P Q R S
T U V W
X Y Z & , .
1 2 3 4 5
6 7 8 9 0

A B C D E F G H

I J K L M N O

P Q R S T U V

X Y Z & , .

1 2 3 4 5 6 7 8 9 0

a b c d e f g h i j

k l m n o p q r s

t u v w x y z

No. 6. CONDENSED DE VINNE.
A B C D E F G H
I J K L M N O
P Q R S T U V
W X Y Z & , .
1 2 3 4 5 6 7 8 9 0
a b c d e f g h i j
k I m n o p q r s
t u v w x y z

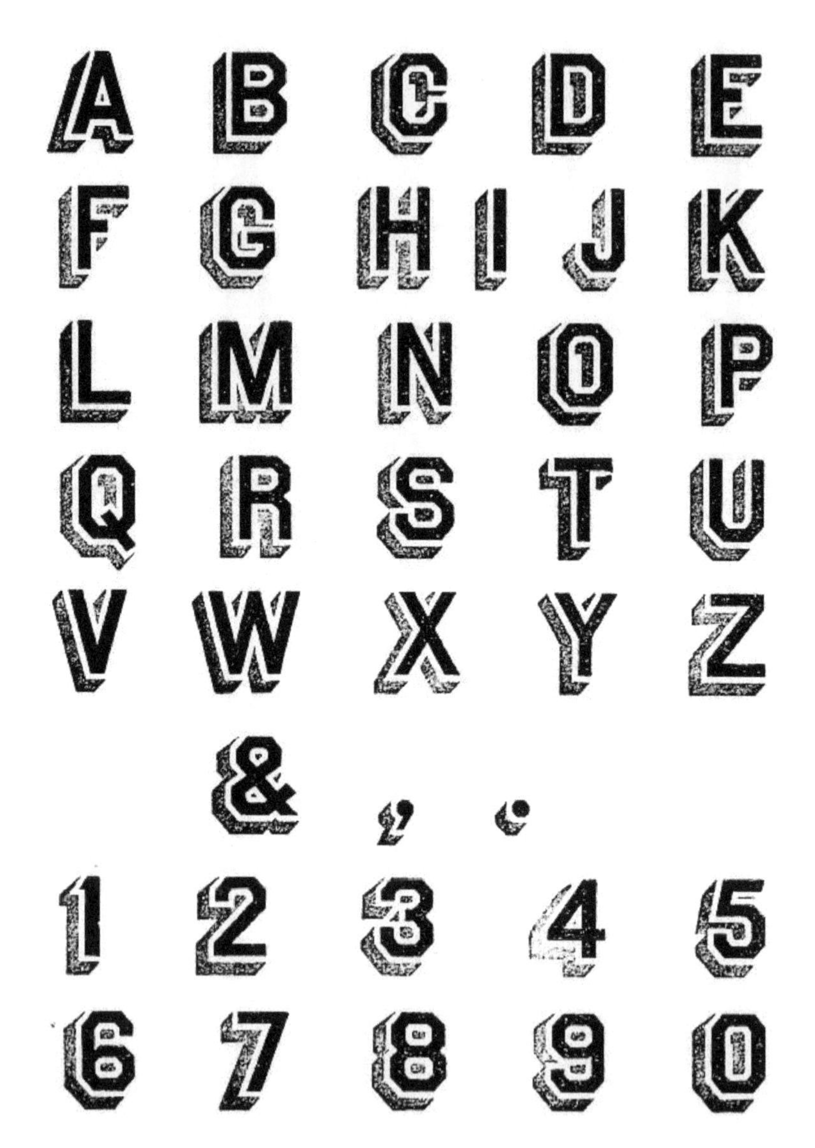

No. 7. GOTHIC SHADED.
A B C D E
F G H I J K
L M N O P
Q R S T U
V W X Y Z
& , .
1 2 3 4 5
6 7 8 9 0

A B C D E F G H I J

K L M N O P Q R S T

U V W X Y Z & , .

1 2 3 4 5 6 7 8 9 0

No. 8. RONALDSON SLOPE.
A B C D E F G H I J
K L M N O P Q R S T
U V W X Y Z & , .
1 2 3 4 5 6 7 8 9 0

A B C D E F G H

I J K L M N O P

Q R S T U V W X

Y Z & , .

1 2 3 4 5 6 7 8 9 0

No. 9. FLORENTINE.
A B C D E F G H
I J K L M N O P
Q R S T U V W X
Y Z & , .
1 2 3 4 5 6 7 8 9 0

A B C D E F G H I
J K L M N O P Q R
S T U V W X Y Z & , .

No. 10. FRENCH OLD STYLE.
A B C D E F G H I
J K L M N O P Q R
S T U V W X Y Z & , .

A B C D E F G H I
J K L M N O P Q
R S T U V W X Y Z
& , . a b c d e f g h i j
k l m n o p q r s t u v
w x y z 1 2 3 4 5 6 7

No. 11. LIVERMORE.
A B C D E F G H I
J K L M N O P Q
R S T U V W X Y Z
& , . a b c d e f g h i j
k l m n o p q r s t u v
w x y z 1 2 3 4 5 6 7
8 9 0

A B C D E F G
H I J K L M N
O P Q R S T U
V W X Y Z & , .
1 2 3 4 5 6 7 8 9
o a b c d e f g h
i j k l m n o p q
r s t u v w x y z

No. 12. CASLON OLD STYLE.
A B C D E F G
H I J K L M N
O P Q R S T U
V W X Y Z & , .
1 2 3 4 5 6 7 8 9
0 a b c d e f g h
i j k l m n o p q
r s t u v w x y z

A B C D E

F G H I J K

L M N O P

Q R S T U

V W X Y Z

No. 13. SATANICK.
A B C D E
F G H I J K
L M N O P
Q R S T U
V W X Y Z

a b c d e f g h

i j k l m n o p

q r s t u v w x

y z 1 2 3 4 5 6

7 8 9 0 . , ! ? &

No. 13. SATANICK—continued.

a b c d e f g h
i j k l m n o p
q r s t u v w x
y z 1 2 3 4 5 6
7 8 9 0 . , ! ? &

A B C D E F G
H I J K L M N O
P Q R S T U V W
X Y Z & , . 1 2 3
4 5 6 7 8 9 0 a
b c d e f g h i j
k l m n o p q r s
t u v w x y z

No. 14. COLUMBUS.
A B C D E F G
H I J K L M N O
P Q R S T U V W
X Y Z & , . 1 2 3
4 5 6 7 8 9 0 a
b c d e f g h i j
k l m n o p q r s
t u v w x y z

A B C D E F G H I J K
L M N O P Q R S T U
V W X Y Z a b c d e f
g h i j k l m n o p q r s
t u v w x y z & , .

No. 15. BRADLEY.
A B C D E F G H I J K
L M N O P Q R S T U
V W X Y Z a b c d e f
g h i j k l m n o p q r s
t u v w x y z & , .

A B C D E F G H I
J K L M N O P Q R
S T U V W X Y Z , .
& 1 2 3 4 5 6 7 8 9 0

No. 16. DORIC ITALIC.
A B C D E F G H I
J K L M N O P Q R
S T U V W X Y Z , .
& 1 2 3 4 5 6 7 8 9 0

LIST OF PRICES AND MODE OF MEASUREMENT.

Prices for Painting and Glazing.

SQUARE MEASURE.

Plain weather boarding, close fencing, ledge doors, partitions, paling fences, etc. All common colors, viz.: White, light yellow, slate, pearl, light drab or cream color, for each coat, per yard	8 cents
Each coat of varnish	10 cents

PANEL WORK.

Flush panel work, panel doors, recesses, etc., the above colors, for each coat, per yard	10 cents
The same in two colors	12 cents
The same in three colors	14 cents
Striping after other work is finished, per foot, lineal measure	1 cent
For expensive or unused colors, per yard, additional	1 cent
For each coat of varnish, per yard	12 cents
For each coat of shellac, per yard	12 cents

BRICK WORK.

	Per Yard.
First coat	15 cents
Second coat	12 cents
Third coat	10 cents
Penciling	15 cents
Mastic or cement, first coat	20 cents

Additional coats, same as brick.

INSIDE WALL PAINTING.

	Per Yard.
First coat	12 cents
Second coat	10 cents
Third coat	8 cents

STOPPING AND CLEANING.

Ordinary puttying, charge price of first coat for the several kinds of work. Puttying longitudinal joints in ceilings, siding, floors, etc., to be charged from two to four times the price of first coat for the several kinds of work, at the discretion of the measurer.

SURFACING, STAINING AND VARNISHING.

Each coat surfacing	10 cents
Each coat stain	8 cents
Each coat varnish	12 cents

LINEAL MEASURE.

Pilasters, architraves, frames, jambs, base mouldings, etc:

	——*Each Coat*——	
Girth.	*Per Foot.*	*Varnish.*
1 to 4 inches	½c	¾c
4 to 6 inches	¾c	1 c
6 to 8 inches	1 c	1¼c
8 to 10 inches	1¼c	1½c
10 to 12 inches	1½c	1¾c
12 to 14 inches	1¾c	2 c
14 to 16 inches	2 c	2¼c
16 to 18 inches	2¼c	2½c
18 to 20 inches	2½c	2¾c
20 to 22 inches	2¾c	3 c

22 to 24 inches	3 c	3¼c

Larger dimensions taken in square measure.

Column mantels as above.

Panel jambs, door casings, etc., to be measured by the above rule.

Plain rosettes, add one foot to length.

Carved rosettes, add two feet to length.

Other carved or ornamental work at the discretion of the measurer.

MODE OF MEASURING.

Begin at wall, press line in all quirks to bead at edge of jamb casing for girth. For jambs take inner sash rabbet to corner bead, double the height and measure between jambs for length.

STRING BOARD, ETC.

	Per Foot.
Plain, each coat	2 cents
Bracketed, each coat	3 cents
Carved, each coat	4 cents
Staff beads, each coat	½ cent
Edge of shelves, each coat	¼ cent

CORNICES AND COLUMNS—PLAIN.

	Per Foot.
Girth, 1 to 2 feet, each coat	3 cents
Girth, 2 to 3 feet, each coat	4 cents
Girth, 3 to 4 feet, each coat	5 cents
Girth, 4 to 5 feet, each coat	6 cents

Plain caps on columns, add to length two feet.

Ornamental caps on columns, add to length four feet.

CORNICES WITH BRACKETS.

	Per Foot.
Girth, 1 to 2 feet, each coat	4 cents
Girth, 2 to 3 feet, each coat	6 cents
Girth, 3 to 4 feet, each coat	8 cents
Girth, 4 to 5 feet, each coat	10 cents
Girth, 5 to 6 feet, each coat	12 cents

Larger dimensions in proportion.

Dental cornices, same price as brackets.

MODE OF MEASURING.

For girth, begin at top, press line into all quirks and over each member to the bottom, and to the length add one-half the medium girth of the brackets multiplied by their number.

PRIMING OR TRACING AND GLAZING SASH.

EACH SIZE, PER LIGHT.

	Priming or Tracing.	*New Glazing.*	*Old Glazing and Glass S.S.*
8 to 10 × 12 to 14	$0.01¼	$0.05	$0.20
12 × 16 or 18	.01½	.08	.35
14 × 24	.02	.10	.40
18 × 24	.03	.14	.50
			D.S.
24 × 30	$ .05	$ .18	$1.00
26 × 36	.06	.20	1.30
30 × 36	.08	.25	1.65

36 × 40	.10	.30
40 × 44	.12	.35
40 × 50	.14	.40
40 × 50	.16	.50
50 × 60	.18	.60
50 × 70	.20	.75

These prices do not apply when called out to glaze one or two lights.

For back puttying add one-quarter, and for bedding add one-half, to the above rates.

In new glazing cost of glass not included.

All breakage at the risk of the owners, if glass is furnished by them. To all bills of glass furnished by the trade 20 per cent. will be charged additional.

PLATE GLASS.

Sizes same as table above, at same prices. Sizes above to 90 square feet, 5 per cent. on net cost delivered; 90 to 108 square feet, 8 per cent.; 108 square feet and upwards, 10 per cent.

Removing old glass, same as above. The owner to pay cost of taking up large glass above first floor.

Unless otherwise provided for, glazier puts glass in at his own risk of breakage, but cutting will be at owner's risk.

SANDING.

First coat of sand equal to two coats of paint, in addition to paint.

Second coat of sand equal to three coats of paint, in addition to paint.

GRAINING—SQUARE MEASURE.

	Per Yard.
Plain oak	$0.40
Plain walnut or ash	.70
Plain satinwood or maple	.70

Plain mahogany or cherry	.70
Shaded oak	.50
Penciled oak or ash	1.00
Penciled chestnut or cherry	1.00
Penciled walnut	1.00
Rosewood	1.00
Oak root	1.50

LINEAL MEASURE.

Girth.	*Graining.*	*Varnishing.*
1 to 4 inches, per foot	\$0.03	\$0.00¾
4 to 6 inches, per foot	.04	.01
6 to 8 inches, per foot	.05	.01¼
8 to 10 inches, per foot	.06	.01½
10 to 12 inches, per foot	.07	.01¾
12 to 14 inches, per foot	.08	.02
14 to 16 inches, per foot	.09	.02¼
16 to 18 inches, per foot	.10	.02½

Other members in proportion.

Graining edges of shelves, per foot, 1½ cents.

Graining sash, double the price of plain painting.

MARBLING—SQUARE MEASURE.

White, per yard	\$0.75
Other kinds, per yard	1.00
Varnishing, each coat, per yard	.12

LINEAL MEASURE.

All members ——*Per foot*——

from	*Marbling.*	*Varnishing.*
1 to 8 inches girth	$0.08	$0.01
8 to 10 inches girth	.12	.01¼
10 to 12 inches girth	.16	.01½
12 to 14 inches girth	.18	.02
14 to 16 inches girth	.20	.02¼

Larger members in proportion.

CLEANING AND KALSOMINING.

Ceilings and walls, per yard	$0.16
Plain cornices, 1 to 2 feet girth, per foot	.02
Plain cornices, 2 to 4 feet girth, per foot	.03

Add to the above for each color, if more than one, 1 cent per foot.

DEDUCTIONS.

The price of any work measured and not specified in this list shall be fixed by the measurer.

The measurer is hereby authorized to deduct from 5 to 20 per cent. from the price of any work that in his judgment is not first-class.

FEES FOR MEASURING.

Jobs amounting to $150 or less	5 per cent.
Jobs amounting to over $150 and less than $500	4 per cent.
Jobs amounting to over $500 and less than $1,000	3 per cent.
Jobs amounting to over $1,000	2 per cent.

Sign Painting.

FACIA SIGNS.

	Gold.	*Plain.*
12 feet long	$ 8.00	$ 4.00
14 feet long	9.00	4.00
16 feet long	10.00	5.00
18 feet long	12.00	6.00
20 feet long	15.00	7.00
24 feet long	16.50	8.00

Above includes two coats of paint.

BRASS SIGNS.

3 × 14 inches	$ 3.50
4 × 20 inches	5.00
6 × 8 inches	4.00
6 × 12 inches	4.50
8 × 14 inches	5.00
10 × 14 inches	5.00
12 × 17 inches	6.00
14 × 20 inches	7.00
18 × 25 inches	10.00
24 × 30 inches	15.00
Sill signs, per square foot	3.50
Square signs, per square foot	3.00

TO MAKE HARD PUTTY.

For Carriage Work.

Mix equal parts of dry __________ and keg white lead with equal parts of rubbing varnish and gold size japan; mix thoroughly and pound well.

For Hurried Work.

Mix dry white lead with equal parts of rubbing varnish and gold size japan. Keep hard putty covered in water when not in use.

TO MAKE AND APPLY KALSOMINE.

Soak one pound good white glue in cold water until soft, then pour off the cold water, and dissolve the glue in hot water. Mix twenty pounds of good whiting in water to a thick paste; dissolve one pound of alum in water, and add it to the mixture. Before mixing the glue and whiting, put in your tinting colors, which should be ground in water. Test your color by dipping in a piecc of paper and letting it dry. After you put in the glue, test in the same way to see if there is enough glue to bind it well, then set your kalsomine aside to get cold.

Thin to good workable consistency with cold water.

Have in enough glue to hold it from washing up when you have to put on a second coat. Too much glue will cause the kalsomine to go on hard, and crack and scale off when dry. If it dries too fast, add two ounces of glycerine to one gallon of kalsomine. Have good staging, and two men for a good sized room. Use good kalsomine brushes, and work fast. Lay on the kalsomine freely; the beauty of the work will depend upon how you lay it off, and level it up. Put it on not as you would paint, *all one way*, but work your brush in all directions, until your work is level, then carefully lay it off with light strokes.

For a white job put in a little blue. If you have never done a job of kalsomining, and have no one to aid you, practice on the wall in your shop or any other place, until you get the knack of it. Cover a small space and see how it comes out.

Always finish lightly with the point of your brush. If an edge dries, stop and wet it with a clean brush and clear water; if careful you can join to it without showing "laps." If you find you have missed any spots wet the edges in the same way, and carefully touch them up with kalsomine. If you find after all your precautions, a water stain has come through your kalsomine, wet the place with a solution of sugar of lead, made in proportion of 1 ounce sugar of lead to 1 quart of rain water; it may kill the stain. See article on wall sizing and water stains, page 39.

Rough places in plaster take more color than a smooth wall, hence they are liable to show spots; so it stands you in hand to make such places smooth as possible; to do this take off the rough sand with sandpaper and knife or trowel on a thin coat of plaster paris, or give the rough places an extra coat or two or size. Fill all cracks and holes, and give the filling time to dry before putting on the size, because otherwise it will take more color than the balance of the wall and your work will look spotted.

In the kalsomining season have some large tubs and mix up as much whiting in hot water as you will need for several days. Add your color, glue, size and alum to *as much only* as you want for immediate use. In hot weather I use liquid glue.

LIQUID GLUE FOR KALSOMINE AND WALL SIZING.

For use in hot weather, a liquid glue which will not decompose and smell badly is very desirable to the workmen and the inmates of the house.

No. 1. To make such a glue fill a bottle a little more than half full of broken up good white glue, and fill the bottle with common whisky or equal parts of alcohol and water. Let it stand a few days and it will dissolve the glue; this glue will keep for years. Keep the bottle corked.

No. 2. Melt your glue in the usual way, thick as you will want it for any purpose, then put in ½ or ¾ ounce *nitric acid* to each pound of glue used; enough to give the glue a sour taste, like vinegar. The acid keeps it in a liquid state, and from spoiling. If you melt the glue in an iron kettle pour it into a wooden vessel, before you add the acid, otherwise the acid will act on the iron and blacken the glue. When wanted for use it can be thinned as desired with cold water; a cask full of this made up and kept air tight so the water will not evaporate will be found very handy to draw from, when you want a little in a hurry for glue size or kalsomine. When you make it up in this way put in at least 1 ounce of acid to the pound of glue to make sure it will keep liquid, so you can draw it from the cask.

Acetic acid will answer the same purpose as nitric acid, but it will take more of it and make the liquid glue more expensive.

TO PREPARE AN OLD WALL FOR PAINT OR PAPER.

First cut out all the cracks V shape, clean out the holes and bevel the edges same as the cracks. Then fill with fine plaster paris mixed with thin glue size. Fill with care; when dry, sandpaper the filling smooth and level. If the wall is sandy or rough, sandpaper it smooth as you can. If the holes are large, have a plasterer stop them, if you can; if you fail in that, and the job must be done soon, fit in thin boards, fill around the edges with plaster, and paste on cloth, or extra paper; but to do a nice job you must insist on having the large holes plastered. If the *hole is up out of reach*, and too large for you to fill, cement the edges with plaster, stretch a piece of cloth, or extra thickness of paper over it, and it will look all right, because the paper will shrink tight when it dries. If you find places where the clinches are broken, and the plaster is loose, press the plaster back to its place if you can, and cut small holes through the plaster and turn small broad headed screws into the lath even with the plaster and cement around the screws with plaster paris.

If it is a smooth wall with rough, sandy patches, sandpaper down the patches a little below the level of the wall, sweep out the loose plaster, give a coat of glue size, and knife or trowel in a coat of plaster paris mixed with glue size or vinegar, and when dry, sandpaper until smooth and level.

There are several points to be considered and provided for in filling cracks in a plastered wall preparatory to painting. First, are the edges of the cracked wall level? To determine this, lay your rule across the crack, and if you find the plaster on one side of the crack higher than the other, it shows that side of the wall has sprung out of place, because the laths are loose or the clinches are broken. The first thing on the program is to get the highest edges back to "place." Failing in that, the next best thing is to raise the other side. If that scheme don't work, the next method is to use sandpaper on a block and rub down the highest side with a wide bevel to match the lowest, otherwise your filling will be at an angle more or less acute with the general surface of the wall, and cast a shadow or reflect the light according to which way the light falls upon it, and the place where the crack was will "show" in spite of your best efforts to conceal it. If you find one edge of a crack higher than the other, gently press against it, and if it goes back to place, cement it with plaster paris wet up in clear water, and it will set in three minutes hard enough to hold the plaster in place. If the loose edge will not go back by gentle pressure, lay a piece of board over it and push hard as you dare to and not crush the plaster. If it is still obstinate, drill out a piece and insert a bent wire or other instrument made on purpose, and see if you can feel the obstruction and remove it. Failing in this, see if you can raise up the lower side to a level with the highest and cement it fast. If the last scheme is too much for your patience and ingenuity, resort to the block and sandpaper, and rub down the high side with a wide bevel to match the other. The next point is to prevent the paint near the edges of the crack, and on the filling which we put in, from drying flat while the balance of the wall bears out a gloss. To do this we must find out the cause of the "flatting" near the edges of the crack and over the "filling." If we examine into the matter, we will find that when the wall cracked the plaster adjacent was more or less fractured and made more porous than the uninjured portions of it. Hence, more oil is drawn from the paint near the crack than where the wall is solid. Now, for the remedy: With a small pointed brush wet the edges of the crack with linseed oil until they will take no more in. Let the oil dry, and fill the crack with plaster mixed with thin glue size, but have the top of the filling one-sixth of an inch below the surface of the wall. Let the filling dry, and with a fine pointed brush paint over the top of the filling and the edges of the crack. Let the paint dry, and finish filling with hard putty. Let the putty dry, and sandpaper the job smooth and level. If you have to bevel the highest edge with sandpaper, first fill the beveled portion with oil. Let the oil dry, and fill the pores with hard putty, because the part beveled with sandpaper will be more porous than the balance of the wall. Treat and fill all small holes by the same method. Filling cracks in this way is a little tedious, I admit; but it is the only way that I know of to stop a crack in plaster, so it will stay stopped and not show after it is painted.

HOW TO PAINT A PLASTERED WALL.

Prime with lead and raw oil, tinted like succeeding coats. Have the prime thin, not more than five pounds of white lead to the gallon of oil; add a little benzine or turps to make it more penetrating. If the room is cool, warm up your prime before you add the benzine or turps. The idea is to have it penetrate as much as possible; brush the prime well into the wall. If it is a sand wall, brush off the loose sand. If it is a smooth one, putty coated or hard finished wall, see that there are no lumps or grains of sand left on the surface. It is a good idea to pass the hand over the wall to feel the lumps, and to knock off lumps and grains of sand by going over the work with sandpaper.

For second coat use glue size, made as directed on another page.

Third coat. Mix so as to dry with a gloss, have the body fairly thick, and spread it well out. Mix with 3 parts linseed oil to 1 part turps.

Fourth coat.—If this coat is to be flat, mix it thick enough to cover well; mix mainly with turps, if the weather is hot, or from any other cause the paint don't work well, add a little linseed oil. For an egg shell gloss, use about 1 part oil *and 3 parts turps.*

If the wall is to be finished in stipple, mix the last coat half oil and half turps, rather thick, and add a little japan. To stipple strike the paint evenly and continuously with the square end of a large brush, made for the purpose; a new clean duster will do. Let the stippler follow the painters. The coat of glue size saves two coats of paint. It is put on after the prime to keep moisture and air from the glue, otherwise it would be liable to decay.

Use boiled oil in all coats except priming coat. Have only enough difference in the color of the different coats, so you can see where you have painted, and not leave holidays; especially in rooms where the light is not very good.

Some painters advocate (especially on hard finished wall) a good filling of clear linseed oil, before any paint is put on to keep the surface from fire cracking.

It is risky business to paint a *new hot* wall; in such cases if it must be done before the lime has become somewhat neutralized, give it a coat of vinegar, and let it stand a day or so before you put on the prime. The vinegar will neutralize the lime and not hurt the priming.

TO PREPARE A ROUGH SANDY WALL FOR PAINT OR PAPER.

If you have a rough brown mortar wall to paper and want to make the job look smooth as possible, first go over it lightly with No. 2 paper to knock off the loose and most prominent grains of sand; then with No. 2 paper rub down all "cat faces" and trowel marks; level up all hollows with plaster paris wet up in thin glue size or vinegar, and you will be ready to put on the lining paper. This paper should be soft and porous so that it will quickly absorb paste and not blister; good white blank wall paper having but little color will answer very well for this purpose. Start in to hang it with half a strip in width so as to break joints with the next coat; use sufficient paste to make the paper stick to the wall; butt the edges and be sure when the paper is dry that there are no loose places. Right here is the turning point of your job for "good or for bad."

Pound the lining paper down so closely that all the prominent grains of sand will show through, and be sure to make it stay there until dry. When the lining paper is dry, go over it with good sharp No. 1½ sandpaper and cut out all the prominent grains of sand which show through the paper, being careful to rub no more than is necessary to take out the sand; the idea being to cut through to the prominently projecting grains of sand, and rattle them out. Some walls will need a second coat of lining paper and another sandpapering, before they are smooth enough for anything like a fine job. If the owner refuses to stand the expense of putting on lining paper, glue size the wall, and when dry, knock off the prominent grains of sand with sandpaper and knife in plaster paris putty on the rough places. In either case, take extra pains with portions of the wall where there are side lights, which always magnify rough places. Sandy walls may be leveled and smoothed somewhat with a coat of kalsomine to hold light bodied paper.

Make a kalsomine of good white glue, 1 pound to 15 pounds of whiting and half a pound of alum. Dissolve the glue and alum in the usual way. When the kalsomine is dry, give the surface a thin coat of glue size to stop the suction. Let the glue size dry, then put on the paper; use light paste, and be sparing of it as you can and make the paper stick. I have often noticed that too much or too little paste is used in paperhanging; some walls and some papers require more paste than others. Too much paste on a smooth wall, or too little on a rough one, makes bad work. If you use a roller for seams have it covered with short plush. To paint on a wall covered with lining paper as above described, first put on a coat of glue size.

TO PAINT OVER NEWLY PLASTERED CRACKS IN WALLS.

When the painter has to paint over holes and cracks in walls recently filled by the plasterer, he will be likely to have to deal with plaster made in part of fresh lime. In such cases, it is always best to soak the newly plastered places *with strong vinegar, to kill as much as possible the caustic properties of the lime.* Put on the vinegar plentifully and let it soak in; when dry, give the new plaster a coat of size made of linseed oil, japan and turpentine; when dry, put on a coat of white shellac before painting.

FLASHED GLASS SIGNS.

Flashed glass is clear on one side and colored on the other; the colored glass forming only a thin film on one side of the clear glass. We can make elegant signs on this glass by etching the letter through the colored portion of the glass, making the letters clear and the background colored; or by etching out the background and leaving the letters colored. Lay out the letters on paper, and place it under the glass as a guide to work by; then, with asphaltum varnish cover the background and leave the letters free and clear; in other words, "cut around them." If you want a clear background with colored border and colored letters, cover the letters and border and leave the background free and clear. Then melt some beeswax, and when it begins to cool, take up a small portion of it with a putty knife and scrape it off on the edge of the glass, and repeat the operation until a wall or dam is made all around the glass, to hold the acid you are about to put on the glass, from running off; then pour on a little hydrofluoric acid, and it will etch out the colored glass not covered by the asphaltum in about one hour; then you can pour the acid back into your bottle, to be used again. Next wash the glass by pouring water over it; then scrape off the wax, and take off the asphaltum with turpentine. Some painters use a varnish made by melting together equal parts of paraffine and asphaltum and thinning to working consistency with turpentine.

FLUORIC ACID, TO MAKE FOR ETCHING PURPOSES.

You can make your own fluoric acid (sometimes called hydrofluoric) by getting the fluor spar, pulverizing it and putting as much of it into sulphuric acid as the acid will cut or dissolve.

Druggists through the country do not keep this acid generally, but they can get it in the principal cities. One ounce will do at least fifty dollars worth of work. It is put in gutta percha bottles or lead bottles, and must be kept in them when not in use, having corks of the same material. Glass, of course, will not hold it, as it dissolves the glass, otherwise it would not etch upon it.

LIQUID WOOD FILLERS FOR CHEAP WORK.

Corn starch and cheap varnish are the principal ingredients of many cheap wood fillers; the corn starch is mixed with the varnish and thinned with turps until workable. *You can experiment on this idea.*

Corn starch in shellac in proportion of 1 pound to the gallon *doubles its capacity as a filler*. I have made and used a filler for cheap work in this way: Pale rosin, 2 pounds; boiled oil, 1 gallon; japan, 1 pint. Melt the rosin in the oil, take the kettle outside, and add ½ gallon turpentine; stir and when cold add ½ pound of corn starch. Thin with turps until workable. Add more or less starch, according to the surface you want to fill. These mixtures are all the better if run through a paint mill.

ANOTHER PASTE FILLER.

Corn starch mixed to a paste with one part linseed oil, two parts each japan and rubbing varnish; thin to working consistency with turpentine.

CARRIAGE PAINTING IN THE VILLAGE SHOP.

NEW WORK.

Prime with white lead, mixed thin in oil, add a little japan and turpentine to make the paint dry hard and quick; when the priming is dry and hard, putty up with hard putty as directed on another page. Then follow with two coats of keg lead thinned with turpentine; add a little japan to make it dry hard, and a little oil to make it work well. Carefully mix and strain your paint. Give the body five coats of rough stuff, made as directed on page 144 and a guide coat, and when dry, proceed to cut down the rough stuff. For this purpose your tools will be several pieces of pumice stone, a pail of water, a large flat file, a good sponge and a chamois. Flatten one side of your stone for a grinding surface and have no thin edges, because they will keep breaking off and be liable to get under the stone, and scratch your work. Now, two of the most important things you will have to guard against is cutting through the rough stuff and lead coats, and scratching the surface. There is a great difference in pieces of pumice stone. Some are hard and full of flint like particles, which will scratch the work; others are softer and of more even grit; the light colored and fairly open grained pieces are the safest to use. You can tell a fast cutting stone by its open grain and lightness. The finer grades of German rubbing brick and English rubbing stone are also used in rubbing rough stuff. A stone with a broad surface is preferable for large surfaces.

Have small pieces to rub around the bolt heads and other places which are difficult to get at with the large stone. The practiced workman can tell the moment a stone begins to scratch, both by the sound and by the feeling to the hand, and you may train your ear and nerve to this degree of sensitiveness; until you do so, you will have to look sharp, and frequently rub your stone on the file, and clean off your work with a sponge full of water to see the condition of the work. Also by passing your hand back and forth across it to determine the condition of it, or if there is any large grit on it, liable to get under the stone and scratch. Rub until the brush marks are gone, etc., which your guide coat will show you. Use plenty of water while rubbing. Thoroughly wash the body inside and out. When dry, sandpaper lightly over the body to remove any grit which may be left on, and to clean out around the irons and panels, also to sand off the irons which you have not rubbed. Dust and wipe well, and when ready, put on a coat of drop black, ground in japan. In mixing your drop black, stir it before you add any turps, then add a little turps, and stir again until it is beaten to a smooth, soft paste; then add sufficient turps to make a workable paint, thin enough to go on easily with a camel hair brush, which for body work on buggies should be not less than one and one-half inches wide and double thick. Painters disagree as to the use of oil in this coat. I like to use a very little good raw oil, say a teaspoonful to a pint of color. It is a good idea to keep a brush on purpose to coat the inside of the body, because it is not usually made as smooth as the outside. Some practice putting on the color coat in the morning and the color varnish towards evening, but I prefer a longer time, say twenty-four hours at least, and more, too, especially when I use a little oil in the color coat. Rub the color with curled hair or hair cloth, dust well, and put on your color varnish; some say with a bristle varnish brush, but I prefer to mix it so I can use a camel hair brush. For this coat mix drop black to a workable paint with equal parts of turps and good body varnish. When this coat is dry, give the body a coat of good rubbing varnish, using a fine bristle varnish brush. Flow on a free coat, lay off to right and left, and finish with up and down strokes across the work. Never put a full brush at the lower edge of the body, because in that case, you will be apt to get a fat edge. Watch for sags or runs, which you can brush out, if discovered before the varnish sets. If a sag or run should get the start of you on this coat, and you see it after the varnish begins to set, squeeze the varnish out of your brush, wet the point of it in turps, and carefully work out the sag or run. Now, dust off the running parts, and put on a coat of color. Some say, have a little more oil in the color for the gear than for the body, but I would not advise the use of more. When dry, put on a coat of color varnish. When dry, rub down with hair or hair cloth, and your gear is ready to stripe.

To paint a wheel, paint one spoke at a time, paint both sides and the edge next to you, then take your brush in your left hand and paint the back edge, and so on, until the spokes are finished. Next paint the hub, then the outside and inside of the felly, then finish the gear, being careful to leave no laps. Use only fine lines for striping a buggy. On the springs, bars, spoke faces, hubs and tongue is all the striping needed. Orange chrome, red, gold, bronze and light green, all harmonize with black, and either may be used for striping a black rig. When ready to varnish, set your gear on trestles. Varnish the wheel with a fine bristle varnish brush, and flow on a full coat. When done with a wheel, set it running on the spindle, and commence the next, and start it off again two or three times, while working at the next wheel, and so on with all the wheels; by this method you may avoid runs, and be able to flow on a fuller coat than you otherwise could. For a finer job, give the gear a coat or two of clear rubbing varnish, and rub each coat down with curled hair or hair cloth. For a cheap job, rub down the body with hair cloth, but for a finer one, rub it out with finely powdered pumice stone in water. For this method, you will need a pail of clear water, some finely powdered pumice stone and a felt pad. The object of this work is to take the gloss off the rubbing varnish, and leave a smooth coat for the finishing varnish. The particular knack is to rub just enough, and then stop; a little too much will cut through, and spoil the job; and not enough will not give you the best possible foundation for your finishing coat of varnish. Keep the work washed off as you go, so you can see defective places, and rub them out. When done rubbing, the next thing is to wash the body perfectly free from grit. Your water brush comes in play here to wash around irons, etc., where the pumice might lodge; then with a pail of clear water, rinse the body and wipe dry with a chamois skin. Right here is a good time to give the inside of the body a coat of color varnish, and to put on your transfers, if you use any. Some painters use a barrel for a body stand, but one made on purpose, of boards, is better. You want to look out for dust in every stage of the work, but right here you must be especially careful, because you are about to put on the finishing coat, which can neither be sandpapered nor rubbed down. You will learn from experience, if not before, that you cannot rely altogether upon the dust brush to free your work from dust and specks. A large soft dry chamois kept for the purpose, and never wet, can be used to advantage to wipe off the dust left by the brush. A hand bellows is very effective in taking dust out of corners where the brush or wiper cannot be worked. When you have done all you can with the brush and wiper, rub the work over with your *hand* and the sensitive nerves of your fingers will detect any specks which may still adhere to the surface. Some other essentials to a good job of varnishing are a clean room, free as possible from dust, clean brushes, and cups, and the person of the varnisher so dressed that he will not shed material for specks. Have one cup to hold your varnish and another to wipe your brush in. Use good varnish and never try to varnish a body with the temperature below 70 degrees F. Have a quill sharpened to a point to pick out any specks which you may discover on your work, because it requires very favorable conditions, and a mighty slick workman to prepare and varnish a body, and not have it show *at least a speck or two.* Use a fine chiseled bristle brush and know that it is absolutely free from specks before you commence. Now, when you are ready, don't be timid or try to see how far you can make your varnish go. Keep in mind from the start that the nearer level—that is, a uniform thickness—you can have your coat of varnish the less liable it will be to sag or run. Put on your varnish with a full brush, laying it on right and left, and brush as level as you can, then finish with up and down strokes, being careful to chisel off the surplus at the lower corner to avoid a flat edge. *Note*—A friend of mine, after laying on his varnish right and left, finished with diagonal strokes across the surface at an angle of 45 degrees, then crossed it again at the same angle in an opposite direction. He had uniform good success.

For an extra fine job, give the work more coats of rubbing varnish, and rub each coat with curled hair, or hair cloth; or you may knife on a coat of putty made of keg lead and equal parts of turps and japan; rub it well in with the flat blade of the knife, and when it sets or flats, scrape off all surplus. Sandpaper when dry. This may go on in the place of third lead. You may, when the job requires it, knife on a coat of hard putty, work it down smooth, let it dry and cut down with sandpaper.

OLD WORK.

There are so many degrees of badness in repair work, that it is not possible to cover the entire ground in a work of this kind. They run all the way from the touch up and varnish job, to the cracked, scaled and almost paintless old rigs. For a touch up and varnish job, at least one which is in decent shape for such work, wash the body, give it a rubbing down with fine powdered pumice stone, clean off and carefully putty cracks, dents, etc., if any; then touch up with color, using a small camel's hair pencil, and cover only where necessary. When dry, give a full coat of body varnish. For a better job, give the body a coat of black rubbing varnish (provided the body is black), then finish with a good coat of wearing body. The gear may be treated the same as the body if in like condition, but if the felloes are worn bare, lead them up and color as you would new work, then touch up the balance and varnish.

The great plague of the paint shop is cracked work, which is otherwise solid. Where the varnish is hard but peeling, take it off with ammonia; to do this, take a side of the body at a time, pour out some ammonia in a cup, and put it on with a clean brush kept for the purpose. Keep the side wet, until you can slice off the varnish with a putty knife; if it fails to come off, you must keep it wet longer. If the varnish is dead and soft, sandpaper down to a solid foundation, then if cracks show sheet up with quick *hard putty* made soft enough to put on with a brush, and scrape off with a knife when set. When dry, sandpaper and if the cracks are not full, give it a second application of putty in the same way. Then for a cheap job give it a coat of color varnish, a coat of rubbing and a coat of body varnish.

If you are to do a fine job, and can get pay for it, and you find the body cracked, burn off the old paint, and commence at the foundation as in new work. For a cheap job, lead up the bare places on the gear and wheels, give a coat of color and a coat of color varnish and finish with heavy gear varnish.

For a fine job, if the old paint is cracked or scaled, take it off and work up from the wood as on a new job.

ROUGH STUFF.

1. To make one coat per day rough stuff, take three pounds of RENO'S filler and one pound of keg lead. Mix to stiff paste with equal parts of rubbing varnish, and first-class japan, thin with turps. Some painters add a little raw oil. Grind the filler fine.

2. French yellow ochre dry, 5 pounds; keg lead 1½ pounds. Mix to stiff paste with equal parts gold size, or best brown japan and rubbing varnish; thin with turps and add a gill of raw oil. *Grind fine.*

CLEANING PHAETON CUSHIONS.

This old phaeton cushion is too dusty for any use, did you say? I agree with you; the old cloth-covered phaeton cushion is one of the unmitigated nuisances which we are often compelled to tolerate in the paint shop. When such a cushion is once filled with dust its capacity for "shedding" seems to be unlimited. The more you beat it and the longer you brush it, the more dust comes to the surface. You can take off a buggy cushion and relegate it to the backroom, but the genius who invented that complicated vehicle called a phaeton, nailed the cushions fast to the body, and we must take them along with the job, dust and all, from the cleaning floor to the varnish room.

When I am so unfortunate as to have an old phaeton brought to my shop, about the first thing I do after cleaning it up is to go for the cushions with the sprinkler and wet them down with clean water, repeating the operation as often as may be necessary to keep in the dust.

Spoil the cushions? No! When you run the rig out of the shop the owner will wonder what you have done to his cushions to make them look so bright. The same operation works well on an old cloth-lined top. After you have brushed all you think you can afford to, and the dust keeps coming to the front, turn the top bottom side up and give it a shower from the sprinkler, and I will guarantee the dust to lie still long enough for you to dress the top and paint the bows. Dust is the natural enemy of the paint shop, and water is one of our best weapons to fight it with.

MIXING QUICK COLOR.

A quick-drying color can be slowed up and made to dry to any required time without injuring it, while if ground in a slow drying preparation, it cannot possibly be quickened without injuring more or less the working and covering properties. The working is certainly important, and the covering more so. The covering property should be strong, because the fewer coats of color on a job the better. Thus a quick dryer saves both labor and time.

Japan colors are best when ground stiff, or with barely enough liquid to bind them firmly, because after being reduced to thinness with turpentine alone they will cling to the surface and will not smut. The color will then have its greatest covering power. Now, by the addition of sufficient pure raw oil to give the best working property, and being also made to dry flat, the color is as near perfection as possible, and the further addition of *anything* weakens the covering power. When an excess of japan is used in grinding, the color is thin, there being less pigment to the pound; and it is of less value to the consumer, while it affords more profit to the manufacturer than when prepared as it should be.

BLACK VARNISH FOR GASOLINE STOVES, ETC.

Asphaltum two pounds, boiled linseed oil one pint, turpentine two quarts. Melt the asphaltum in an iron pot, heat the oil, and add it to the asphaltum while hot. Stir well. When partly cool, add the turpentine and four ounces of good japan.

BLACK STENCH INK.

Shellac two ounces, borax two ounces, soft water twenty ounces, gum arabic two ounces, lampblack and indigo sufficient. Boil the shellac and borax in the water until dissolved, then add the gum arabic; dissolve and take the mixture from the fire; when cold, add enough lampblack to give it color and proper consistency, and a little powdered indigo. Keep in glass or earthenware vessels.

BRONZE FOR BRIGHT METALS.

Red aniline (fuscine) ten parts, purple aniline five parts, alcohol 95 per cent. one hundred parts, benzoic acid five parts. Add the anilines to the alcohol, and dissolve by placing the vessel in a sand or water bath. As soon as dissolved, add the benzoic acid and boil for five or ten minutes, or until the greenish color of the mixture is turned to a brilliant light bronze; spread with a brush on bright metal.

VARNISH TO FIX PENCIL DRAWINGS.

Gum mastic three ounces, alcohol one pint. Dissolve and apply with a brush.

RUST SPOTS ON MARBLE.

Apply a mixture of 1 part nitric acid and 25 parts of water, then rinse with 3 parts water and 1 part ammonia.

WHITEWASH TO SOFTEN.

To soften old whitewash which you wish to take off, wet it thoroughly with a wash made of 1 pound of potash, dissolved in 10 quarts of water.

WATER GLASS FOR FLOORS.

Clean the floor, fill cracks with water glass cement made of water glass and whiting, then put on a coat of water glass, to be followed by second coat; when dry rub the last coat with pumice stone and oil.

TO FINISH REDWOOD.

Take one quart of spirits turpentine; add one pound of corn starch; quarter of a pound burnt sienna; one tablespoonful raw linseed oil and one tablespoonful brown japan. Mix thoroughly, apply with the brush, let it stand, say, fifteen minutes, rub off all you can with fine shavings or a soft rag, let it stand at least twenty-four hours that it may sink into and harden the fibers of the wood; afterward apply two coats of white shellac, rub down well with fine flint paper, then put on from two to five coats best polishing varnish; after it is well dried rub with water and pumice stone ground very fine; stand a day to dry; after being washed clean with a chamois rub with water and rotten stone; dry; wash as before clean, and rub with olive oil until dry. Some use cork for sandpapering and polishing, but a smooth block of hardwood like maple is better. When treated in this way, redwood will be found the peer of any wood for real beauty and life as a house trim or finish.

MARKING INK.

Asphaltum, dissolved in turpentine to a thin fluid, will give you an excellent marking ink for all purposes; dries quickly, does not spread, and is nearly indestructible.

FORMULAS FOR MIXING COLORS. (SELECTED.)

It is impossible to give infallible recipes for mixing colors, on account of the difference in the tone and color strength of pigments, both dry and in oil, many samples having as high as fifty per cent. of barytes or other white makewright material, which not only lessens the color strength of the mixture in proportion to their volume, but weakens the color, in a small measure, by their presence as white material. Hence, color formulas are made subject to modification, not only to please the taste of the mixer, but on account of the presence of poor, weak and adulterated pigments.

The writer has selected a few formulas from which the learner may gain some knowledge of colors, which he can improve upon by experiment.

NOTE.—Part means in bulk, not by weight.

Plumb.—White lead 2 parts; Indian red, 1 part; ultramarine blue, 1 part. If too dark, add more white lead. (Outside.)

Brick.—Yellow ochre, 2 parts; Venetian red, 1 part; white lead, 1 part. If too dark, add more ochre. Don't depend upon the common ochre of the stores. It has but little tinting power. Use French ochre ground in oil. (Outside.)

Bronze Green.—Chrome green, 5 parts; lampblack, 1 part; burnt umber, 1 part. If too dark, use more green. (Outside.)

Jonquil Yellow.—White lead tinted with chrome yellow and vermilion.

Lead Color.—White Lead, 16 parts; ultramarine blue, 1 part; lampblack, 2 parts. (Outside.)

Light Buff.—White lead tinted with yellow ochre (Outside.)

Lemon.—Lemon chrome yellow, 5 parts; white lead, 2 parts. (Outside.)

Brown.—Indian red, 3 parts; lamp black, 2 parts; yellow ochre, 1 part. If too dark, use more ochre or less black. (Outside.)

Chestnut.—Venetian red, 2 parts; lamp black, 1 part; medium chrome yellow, 4 parts. (Outside.)

Lilac.—Light Indian red, 3 parts; white lead, 3 parts; ultramarine blue, 1 part.

Purple.—Light Indian red, 4 parts; white lead, 3 parts; ultramarine blue, 2 parts.

London Smoke.—Burnt umber, 2 parts; white lead, 1 part; Venetian red, 1 part.

Brown.—Venetian red, 3 parts; drop black, 2 parts; chrome yellow, 1 part. (Outside.)

French Gray.—White, tinted with ivory or drop black. (Outside.)

Olive Yellow.—Burnt umber, 3 parts; lemon chrome yellow, 1 part. For lighter shade, add more yellow.

Pearl.—White lead, 6 parts; Venetian red, 2 parts; lamp black, 1 part. If too dark, add more lead. (Outside.)

Olive.—Lemon chrome yellow, 10 parts; ultramarine blue, 1 part; light Indian red, 1 part.

Cream Color.—White lead, 8 parts; French yellow ochre in oil, 2 parts; Venetian red, 1 part. (Outside.)

Tan.—Burnt sienna, 5 parts; medium chrome yellow, 2 parts; raw umber, 1 part. If too red, add more raw umber.

Pea Green.—White lead, 5 parts; chrome green, 1 part. Vary the proportions to suit.

Drab.—White lead, 10 parts; burnt umber, 1 part. Vary to suit.

Canary.—White lead, 6 parts; lemon chrome yellow, 2 parts, or less, as you like it. (Outside.)

Fawn.—White lead, 8 parts; chrome yellow, 1 part; Indian red, 1 part; burnt umber, 1 part. (Outside.)

Grass Green.—Lemon chrome yellow, 3 parts; Prussian blue, 1 part.

Peach Blossom.—White lead, 1 part; light Indian red, 1 part; ultramarine blue, 1 part; lemon chrome yellow, 1 part.

Light Gray.—White lead, 10 parts; ultramarine blue, 1 part; lampblack, 1 part. Make lighter or darker by using more or less white lead, as the case may require.

Purple Brown.—Dark Indian red, 4 parts; ultramarine blue, 1 part; lampblack, 1 part. Light up with white lead to fancy. If too purple, use less blue; if too red, use more black. (Outside.)

Leather Brown.—Venetian red, 2 parts; yellow ochre, 4 parts; lampblack, 1 part; white lead, 2 parts or more, to suit. If too dark, use less black. (Outside.)

Dregs of Wine.—Tuscan red with a little lampblack and white lead.

Leaf Bud.—Equal parts white lead, orange chrome and chrome green. If too dark, add more lead. (Inside only.)

Coral Pink.—Vermilion (English), 5 parts; white lead, 2 parts; chrome yellow, 1 part. (Inside.)

Maroon.—Tuscan red, 3 parts; ultramarine blue, 1 part. If too red, add more blue.

Myrtle.—Dark chrome green, 3 parts; ultramarine blue, 1 part. Light up with white lead.

Stone.—White lead, 5 parts; French yellow ochre, 2 parts; burnt umber, 1 part. Tint to desired shade with raw umber; a very little will do. (Outside.)

Snuff.—Medium chrome yellow, 4 parts; Vandyke brown, 2 parts.

Rose.—White lead, 5 parts; carmine, 2 parts. (Inside only.)

Portland Stone.—Raw umber, 3 parts; yellow ochre, 3 parts; white lead, 1 part. (Outside.)

Ashes of Roses.—White, lightly tinted with black, blue and lake. (Inside only.)

Silver Gray.—Tint white lead with lampblack and indigo.

Fine Chocolate.—Tint the best burnt umber with Munich lake. (Inside only.)

Fine Maroon.—Tint any deep red lake with a little orange chrome yellow.

Vienna Smoke.—Tint fine burnt umber with lemon chrome yellow and a little Venetian red.

Quaker Green.—Chrome green, 3 parts; lampblack, 1 part; Venetian red, 1 part; medium chrome yellow, 1 part.

Chamoline.—Lemon yellow, 1 part; raw sienna, 3 parts; white lead, 5 parts.

Clay Drab.—White lead, raw sienna, raw umber, equal parts. Tint with chrome green.

Pearl.—White lead, tinted with ultramarine blue and lampblack.

Copper.—Medium chrome yellow, 2 parts; Venetian red, 1 part; drop black, 1 part.

Buttercup.—White lead tinted with lemon chrome yellow.

Flesh.—White lead, 8 parts; light Venetian red, 1 part; orange chrome, 2 parts.

Olive Brown.—Lemon chrome yellow, 1 part; burnt umber, 3 parts.

Deep Buff.—White lead tinted with yellow ochre and a little Venetian red. (Outside.)

SOME EXPENSIVE COLORS.

Claret.—Carmine, 2 parts; ultramarine blue, 1 part.

Carnation Red.—Carmine lake, 3 parts; white lead, 1 part.

Chocolate.—Fine burnt umber, 5 parts; carmine or lake, 1 part.

French Red.—Indian red and English vermilion, equal parts, glazed with carmine.

Rose.—White lead, 5 parts; carmine, 2 parts.

Yellow Lake.—Burnt umber and white lead, equal parts; tint with chrome yellow and lake.

SUGGESTIONS FOR TINTS AND COLORS.

Delicate Flesh Tints, white predominating.—1st, white and light red; 2nd, white, Naples yellow, vermilion; 3rd, white, vermilion and light red.

Gray and Half Tints, white predominating.—1st, white, vermilion and black; 2nd, white and terre verte; 3rd, white, black, Indian red and raw umber.

Deep Shades, color predominating.—1st, light red and raw umber; 2nd, Indian red, lake and black.

Carnations.—1st, white and Indian red; 2nd, white and rose madder; 3rd, white and lake; 4th, white and Naples yellow.

Carnations, color predominating.—1st, rose madder and white; 2nd, Indian red, rose madder and white.

Green Tints.—1st, white and ultramarine blue, with any yellow; 2nd, white and terre verte; add a little raw umber.

Gray Tints.—1st, ultramarine blue, light red and white; 2nd, Indian red lake, black and white.

Pearly White, white predominating.—1st, white, vermilion and black; 2nd, white, vermilion and black; 3rd, white and black.

Gray.—White, Venetian red and black.

Yellow.—Yellow ochre and white.

Olive.—Yellow ochre, terre verte and umber.

Sky.—French blue and white.

PAINTING CARS AT HOME.

Probably no other subject dealing with the problem of the motorist has been so little, or to be correct, so unsatisfactorily treated as the home painting of cars. Most of the literature dealing with the subject is written in a technical vein, purely for the delectation of the professional painter. This naturally leads the novice to believe, owing to the great number of coats these writers say is essential for good work, that it is entirely out of the question for a car owner, without previous experience in painting, to repaint his car satisfactorily.

Fortunately this is not true. In the first place, the fewer number of coats that can be applied and still accomplish the desired result, will make far the most durable and lasting job of painting. I maintain, and have proven, times without number, that if a motorist really is in earnest about wanting to paint his cars, the battle is more than half won. Give this class of motorists the proper material mixed ready for use with the proper brushes for their application, and tell him how they should be used, and 99 times out of 100 he will paint his car so well that he will be sorry he had not done it before.

The quality of the material used in this kind of painting is of vital importance and unless they are the very best will give but limited wear. And the proper brushes to use with the different coats is of equal importance. About four-fifths the cost of automobile painting is labor, so that a few dollars saved in buying the materials is false economy.

The general purpose enamels for sale in stores have no place on a motor car. They belong to the home. Probably no other vehicle excepting a locomotive has harder service for paint to withstand. Hence, the necessity for the very best materials.

There is one reputable concern selling repainting outfits to car owners so that greatly simplifies the painting problem, if one wishes to do it himself. These outfits include everything, materials, brushes, and instructions, and range in price from $6 for a small runabout to $8 and $10 for a roadster and touring car. Compared to $35 to upwards of $100 that one has to pay a regular painter, if one wishes to economize, the painting affords a grand opportunity.

The fenders and hood of a car are subjected to severe wear and the time is coming when these parts will always be painted black, regardless of the body color of the car. In fact, a great many of the new cars will be painted this way. There are thousands of cars in use that hardly need repainting, but if the hood and fenders were done over in black it would make them look almost like new cars. There is one concern making these hood and fender outfits and a novice can do a really creditable job of painting with them. They range in price from $3 to $5 and are in two coats with a suitable brush.

The gases from the motor are a big factor in dulling the paint on hoods. It has the same effect that ammonia fumes from a stable has on horse-drawn vehicles. This is one reason why the black painting of hoods is mighty sensible. When your hood gets dull, instead of laying up your car you can paint the hood yourself with little trouble and no loss of time.

For the novice to repaint the average car, for instance a 30 H. P. touring car, it would require in labor only a few hours on four or five different days. The hardest part of the whole operation is preparing the car for paint. It is absolutely necessary to have it thoroughly clean before applying any paint. It should be well washed first, and then given a gasoline bath to the parts on which dirt and grease have been allowed to accumulate. It is really not so complex a proposition after all. If a woman can paint furniture with enamels that are no better than they should be, a man can surely paint a car if given the proper materials to do it with, and if he be instructed in their use.

Now as to striping. This is of course out of the question for the novice. But you can black the mouldings of the body, seats, doors, hubs and rims of wheels so that the absence of striping is not noticed. So far as the striping goes, the tendency is away from it—in fact, the most expensive cars have hardly any striping. The blacking of the mouldings, etc., mentioned makes a harmonious contrast and takes the place of striping. It looks in no ways amateurish—rather like the handiwork of the professional painter.

In addition to the saving that can be effected by repainting your car yourself, there is the feeling of personal pride when the job is finished, of having done something well yourself.

As the majority of the new cars have enameled lamps instead of polished brass as in years past, I believe a few words on the subject will not be amiss. In my experience of twenty years in the painting of vehicles, locomotives and automobiles, I have never had a harder proposition to solve than the enameling of polished brass lamps, particularly gas headlights.

An enamel for this purpose must of necessity be made highly elastic, so that it will contract and expand with the metal and stick on the polished brass surface without any previous roughing. This means that only the most expensive materials can be used in the making of such an enamel. There is one enamel of proven merit for this purpose on the market and it does not have to be baked. I have seen a great many motorists who have used general purpose enamels on their lamps and the experience has usually been that the enamel leaves when the lights are lighted. If I were buying an enamel for use on the brass parts of my car, I should be very careful to buy the one that had been long on the market, for there will undoubtedly be a large number of new ones offered.

I have made some pretty strong statements in the foregoing article, and it is no more than right that I tell you that they are based on my experience of twenty years in the painting of carriages, locomotives and automobiles, two years as the expert for the largest paint and color house in the world, and several years in the manufacture of the highest class of motor car paints.

Milton Keynes UK
Ingram Content Group UK Ltd.
UKHW012253110624
443988UK00006B/393

9 789361 470813